NUCLEAR MATERIALS AND DISASTER RESEARCH

PROTECTION OF RADIOACTIVE MATERIAL

NUCLEAR REGULATORY COMMISSION RULE 10 C.F.R. 37

NUCLEAR MATERIALS AND DISASTER RESEARCH

Additional books in this series can be found on Nova's website under the Series tab.

Additional E-books in this series can be found on Nova's website under the E-book tab.

NUCLEAR MATERIALS AND DISASTER RESEARCH

PROTECTION OF RADIOACTIVE MATERIAL

NUCLEAR REGULATORY COMMISSION RULE 10 C.F.R. 37

ELIJAH B. EVANS
AND
WYATT C. JOHNSON
EDITORS

nova publishers
New York

For permission to use material from this book please contact us:
Telephone 631-231-7269; Fax 631-231-8175
Web Site: http://www.novapublishers.com

Library of Congress Cataloging-in-Publication Data

ISBN: 978-1-62808-563-1

Published by Nova Science Publishers, Inc. † New York

CONTENTS

PREFACE

This book analyzes 10 C.F.R. 37, a new rule promulgated by the Nuclear Regulatory Commission (NRC) to protect radioactive material. "Byproduct material" includes specified types of radioactive material other than uranium or plutonium. The rule regulates byproduct material of types and in quantities that could be used to make a "dirty bomb." Congress attaches great importance to protecting the United States against terrorist threats and this rule will affect the many industrial, research, and medical activities nationwide that use radioactive materials, thereby affecting many constituents and raising cost-benefit issues. There is wide concern also about regulation and radiation.

Chapter 1 – Congress may find the above analysis of interest for several reasons: Congress attaches great importance to protecting the United States against terrorist threats; the rule will affect the many industrial, research, and medical activities nationwide that use radioactive materials, thereby affecting many constituents and raising cost-benefit issues; and there is wide concern about regulation and radiation more generally.

NRC may regulate through orders or rules. Both have the effect of law. If prompt action is required, NRC may issue orders to its licensees. After 9/11, NRC issued orders to enhance radioactive material security. Orders went to licensees of irradiators having a large amount of radioactive material (2003), manufacturers and distributors of radioactive material (2004), licensees transporting radioactive materials in quantities of concern (2005), and others.

NRC prefers, however, to regulate through rules. The rulemaking process is transparent, as it solicits public comments and revises draft rules to reflect them, a process that may take years, whereas NRC may issue orders with little or no public involvement. An order applies only to licensees receiving it, while a rule applies to all current and future licensees. Accordingly, NRC began a rulemaking process to consolidate the security orders into a rule. It solicited public comments in the June 15, 2010, *Federal Register.*

10 C.F.R. 37 incorporates many provisions of the orders. Its main areas are:

- Background investigations and access control programs, setting trustworthiness and reliability (T&R) requirements for persons granted unescorted access to radioactive material in quantities of concern.
- Physical protection requirements during use, requiring licensees to establish a written security program, coordinate with local law enforcement, and be able to monitor, detect, and assess theft of radioactive material.

- Physical protection in transit, requiring transporters of radioactive material to follow certain procedures.

To assess impacts of the rule, CRS interviewed 14 radiation professionals, including state regulators, university radiation safety officers, manufacturers and distributors of radioactive material, and a transportation specialist. They noted such impacts as:

- T&R adjudication: Some interviewees felt requirements were burdensome; others could meet requirements easily. Some felt human resources staff lacked expertise needed to adjudicate T&R; others were confident in staff capability.
- Cooperation with local law enforcement: Some interviewees felt that cooperation was excellent. Some found it improved greatly once the threat had been explained. Others found poor cooperation. Still others noted that multiple emergencies or the size of an officer's patrol area could delay response.
- Transportation security: NRC relinquishes parts of its regulatory authority to certain states. One interviewee felt the Department of Transportation, not NRC and states, should regulate transport of radioactive materials to establish uniform national requirements: "when transportation requirements are set by many [state] regulators and by federal interstate commerce rules as well, compliance with transportation regulations can become very complicated."

The analysis raises several general points:

- The orders imposed many requirements on licensees. Since the rule incorporates many of these requirements, it imposes less additional regulatory burden than would have been the case if the orders had not been issued first.
- Since the rule requires many actions regardless of facility size, it would appear to impose a proportionately larger burden on small licensees.
- The rule implements a layered defense. The ability of one layer to offset weaknesses in others should improve security but cannot guarantee it.
- Further steps—such as requiring licensees to install more monitoring equipment or mandating more stringent requirements for adjudicating personnel reliability— might increase security beyond what the rule provides. Whether their costs are worth potential benefits is a matter for political judgment.

Chapter 2 – The intent of this technical report is to provide guidance on, and to assist applicants and licensees in, the implementation of Title 10 of the *Code of Federal Regulations* (10 CFR) Part 37, "Physical Protection of Category 1 and Category 2 Quantities of Radioactive Material." This document describes methods that the U.S. Nuclear Regulatory Commission (NRC) finds acceptable for implementing the regulations.

In: Protection of Radioactive Material
Editors: Elijah B. Evans and Wyatt C. Johnson
ISBN: 978-1-62808-563-1

Chapter 1

NUCLEAR REGULATORY COMMISSION 10 C.F.R. 37, A NEW RULE TO PROTECT RADIOACTIVE MATERIAL: BACKGROUND, SUMMARY, VIEWS FROM THE FIELD*

Jonathan Medalia

SUMMARY

This report analyzes 10 C.F.R. 37, a rule promulgated by the Nuclear Regulatory Commission (NRC), "Physical Protection of Byproduct Material." "Byproduct material" includes specified types of radioactive material other than uranium or plutonium. The rule regulates byproduct material of types and in quantities that could be used to make a "dirty bomb."

Congress may find this analysis of interest for several reasons: Congress attaches great importance to protecting the United States against terrorist threats; the rule will affect the many industrial, research, and medical activities nationwide that use radioactive materials, thereby affecting many constituents and raising cost-benefit issues; and there is wide concern about regulation and radiation more generally.

NRC may regulate through orders or rules. Both have the effect of law. If prompt action is required, NRC may issue orders to its licensees. After 9/11, NRC issued orders to enhance radioactive material security. Orders went to licensees of irradiators having a large amount of radioactive material (2003), manufacturers and distributors of radioactive material (2004), licensees transporting radioactive materials in quantities of concern (2005), and others.

NRC prefers, however, to regulate through rules. The rulemaking process is transparent, as it solicits public comments and revises draft rules to reflect them, a process that may take years, whereas NRC may issue orders with little or no public involvement. An order applies only to licensees receiving it, while a rule applies to all current and future licensees. Accordingly, NRC began a rulemaking process to

---

* This is an edited, reformatted and augmented version of Congressional Research Service, Publication No. R42868, dated December 14, 2012.

consolidate the security orders into a rule. It solicited public comments in the June 15, 2010, *Federal Register.*

10 C.F.R. 37 incorporates many provisions of the orders. Its main areas are:

- Background investigations and access control programs, setting trustworthiness and reliability (T&R) requirements for persons granted unescorted access to radioactive material in quantities of concern.
- Physical protection requirements during use, requiring licensees to establish a written security program, coordinate with local law enforcement, and be able to monitor, detect, and assess theft of radioactive material.
- Physical protection in transit, requiring transporters of radioactive material to follow certain procedures.

To assess impacts of the rule, CRS interviewed 14 radiation professionals, including state regulators, university radiation safety officers, manufacturers and distributors of radioactive material, and a transportation specialist. They noted such impacts as:

- T&R adjudication: Some interviewees felt requirements were burdensome; others could meet requirements easily. Some felt human resources staff lacked expertise needed to adjudicate T&R; others were confident in staff capability.
- Cooperation with local law enforcement: Some interviewees felt that cooperation was excellent. Some found it improved greatly once the threat had been explained. Others found poor cooperation. Still others noted that multiple emergencies or the size of an officer's patrol area could delay response.
- Transportation security: NRC relinquishes parts of its regulatory authority to certain states. One interviewee felt the Department of Transportation, not NRC and states, should regulate transport of radioactive materials to establish uniform national requirements: "when transportation requirements are set by many [state] regulators and by federal interstate commerce rules as well, compliance with transportation regulations can become very complicated."

The analysis raises several general points:

- The orders imposed many requirements on licensees. Since the rule incorporates many of these requirements, it imposes less additional regulatory burden than would have been the case if the orders had not been issued first.
- Since the rule requires many actions regardless of facility size, it would appear to impose a proportionately larger burden on small licensees.
- The rule implements a layered defense. The ability of one layer to offset weaknesses in others should improve security but cannot guarantee it.
- Further steps—such as requiring licensees to install more monitoring equipment or mandating more stringent requirements for adjudicating personnel reliability— might increase security beyond what the rule provides. Whether their costs are worth potential benefits is a matter for political judgment.

INTRODUCTION

A "dirty bomb" or other type of radiological dispersal device (RDD, a device for dispersing radioactive material) could contaminate several square miles of densely populated or economically critical areas, such as a key port or the center of a large city. (The amount of area contaminated would depend on many factors.) The types of radioactive material most likely to be used in an RDD are widely available. While radioactive contamination from an RDD would kill few people, many experts believe, it could render an area off-limits, perhaps for many years. That, in turn, would require a massive cleanup effort, possibly including demolition and reconstruction of buildings and streets. As a result, an RDD attack could impose tens of billions of dollars of costs from economic activity forgone, societal disruption, and cleanup and remediation.

Congress has acted in multiple ways to protect the United States against the threat of terrorists seizing radioactive material that they would need to make an RDD:

- Congress has held numerous hearings on securing and detecting nuclear and other radioactive materials, and the threat they pose.[1]
- The SAFE Port Act of 2006 (P.L. 109-347), Section 121 (6 USC 921) mandated that all cargo containers entering the United States through certain ports "shall be scanned for radiation. To the extent practicable, the Secretary [of Homeland Security] shall deploy next generation radiation detection technology."
- Section 501 (6 USC 591) of that Act established the Domestic Nuclear Detection Office and made it "responsible for coordinating Federal efforts to detect and protect against the unauthorized importation, possession, storage, transportation, development, or use of a nuclear explosive device, fissile material, or radiological material in the United States, and to protect against attack using such devices or materials."
- Congress has funded the Global Threat Reduction Initiative of the National Nuclear Security Administration (NNSA). Among other things, this initiative secures nuclear and other radioactive material domestically and internationally.
- Congress has funded programs by the Departments of Homeland Security, Defense, State, and Energy to research, develop, procure, and operate equipment to detect nuclear and other radioactive material at U.S. and foreign ports and elsewhere.

The Nuclear Regulatory Commission (NRC) is one of several agencies playing a role in protecting the United States against an RDD attack. NRC is in charge of regulating civilian nuclear and other radioactive materials. In the wake of 9/11, it issued a series of orders to its licensees mandating steps to increase security of many types of radioactive materials. Orders are effective immediately, issued with little or no formal input from stakeholders, and apply only to licensees to whom they are issued. Because of these limitations, NRC has prepared a new rule that it has developed with extensive input from stakeholders and that would apply to all licensees. This rule, 10 C.F.R. 37, is "Physical Protection of Byproduct Material." ("Byproduct material" includes specified types radioactive material other than uranium or plutonium as defined in **Appendix A**.) Combining features of the security orders, the rule would regulate radioactive material of the types and quantities that pose the greatest threat for use in an RDD. The rule will have broad impacts across the country and across most if not all aspects of industries that use radioactive material, including hospital and blood bank irradiators, industrial radiography equipment, massive facilities for irradiating certain foods and medical supplies, laboratory equipment for research into radiation and its effects, state regulators, and manufacturers, distributors, and transporters of radioactive sources. NRC anticipates that the rule will be published in the *Federal* Register in early 2013. It would be effective one to several years after publication; until then, the orders would remain in effect.

Congress may find this analysis of interest for several reasons: Congress attaches great importance to protecting the United States against terrorist threats; the rule will affect the many industrial, research, and medical activities nationwide that use radioactive materials, thereby affecting many constituents and raising cost-benefit issues; and there is wide concern about regulation and radiation more generally. This report analyzes the rule.

In order to analyze the rule, CRS conducted detailed telephone interviews with 14 professionals in the radiation industry.[2] They include radiation safety officers—individuals responsible for the safety and security of radioactive material and radiation equipment—at large universities; industrial radiographers; state regulators; manufacturers and distributors of radioactive sources; a representative of the Nuclear Energy Institute, which describes itself as "the policy organization of the nuclear energy and technologies industry";[3] and a transporter of radioactive material. These interviews—which provided the views of front-line radiation professionals—constitute the database for this analysis of 10 C.F.R. 37. They display the widespread effects of this new rule, and the widely divergent views that these professionals hold of virtually every part of this rule. At the same time, one interviewee commented, "Part 37 is a sweeping, high-impact rule that affects the entire industry"; based on their responses, most if not all interviewees would seem to agree.

TECHNICAL BACKGROUND

While there had been concerns about the vulnerability of the United States to "weapons of mass destruction," or more precisely chemical, biological, radiological, and nuclear (CBRN) weapons, for many years prior to September 11, 2001, the events of that day elevated concern about CBRN attacks, especially those that might be mounted by "non-state actors" such as terrorists.

One concern is that a terrorist group might acquire and set off a radiological dispersal device, or RDD. An RDD might contaminate several square miles with radioactive material. Such material would probably result in few prompt fatalities, but could cause additional long-term cancer fatalities. Its main potential impact would be rendering key sections of a city or critical infrastructure (e.g., Wall Street, Pennsylvania Avenue, the Port of Los Angeles-Long Beach) off-limits for years, causing immense societal and economic disruption, forcing many people to relocate, and requiring cleanup that could cost billions of dollars. A "dirty bomb," in which radioactive material is dispersed using explosives, is one type of RDD, but other means of dispersal are possible, such as dropping material from a tall building on a windy day.

Every chemical element has at least one radionuclide, or radioactive form of its atom, but only a few are of concern for an RDD. Radionuclides with half-lives of a few days or less decay quickly, reducing cleanup costs and societal dislocation. Radionuclides with half-lives of tens of thousands of years emit so little radiation per unit mass that tons might be required to emit as much radiation, per unit time, as a few grams of other materials. Different types of radiation pose different threats. Most radionuclides are rare or nonexistent in commerce. The International Atomic Energy Agency (IAEA) screened radionuclides and found 16 of particular concern.[4] **Appendix B** shows these radionuclides. In 2004, the IAEA produced a Code of Conduct on the Safety and Security of Radioactive Sources that was intended to "serve as guidance to States for—*inter alia*—the development and harmonization of policies, laws and regulations on the safety and security of radioactive sources."[5] It applies to specified quantities of the 16 radionuclides. A category 1 source of a radionuclide has 100 times as much material as a category 2 source, which has 10 times as much material as a category 3 source. Most category 2 quantities of the 16 radionuclides weigh a few grams or less. While the IAEA defines these categories in terms of harm to an individual, they have become the international standard for defining harmful quantities of materials meriting protection against possible terrorist use. 10 C.F.R. 37 applies to the IAEA's category 1 and 2 quantities of the 16 specified radionuclides.[6]

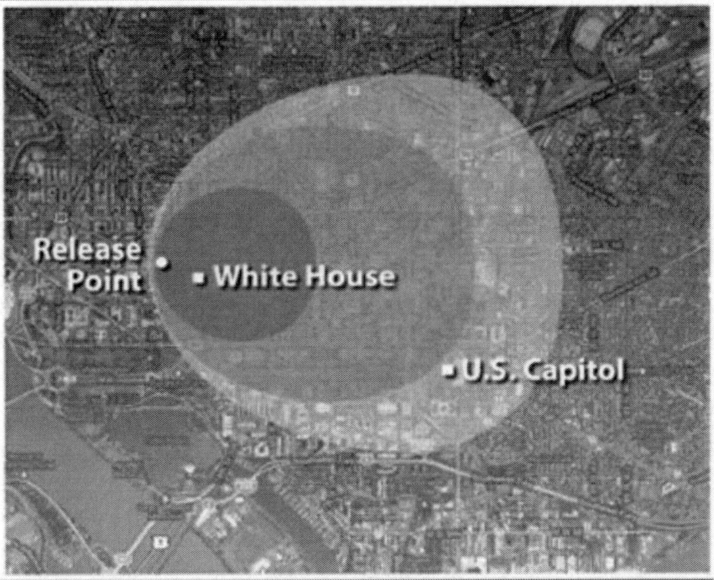

Effects and Actions

	Area km² mi	Equivalent Dose (rem)	Exceeds relocation PAG for which year:	Population	All Cancers	Fatal Cancers
	2.10 0.81	>2.00	First year only	38,000	233	159
	7.60 2.93	>0.500	Any subsequent year	94,700	278	189
	13.2 5.10	>5.00	50 years (cumulative)	125,000	461	314

Areas and counts are cumulative. RDD detonated at 38.9 N. 77.0 W. PAG: Protective Action Guide

Source: William Rhodes III, Senior Manager, International Security Systems Group, Sandia National Laboratories, September 2010; analysis by Heather Pennington; graphics by Mona Aragon.

Notes (provided by William Rhodes): This map, based on an atmospheric dispersion model, shows where individuals are projected to have an increased risk of developing cancers due to radiation exposure over a year or more. The RDD in this scenario uses 1,000 curies of cesium-137 chloride (about 50 grams). The model assumes that all material used is dispersed, but that it is not dispersed evenly over the area. Wind is assumed to be from west to east at 7 mph. The model includes exposure from radioactive material both deposited on the surface and resuspended into the air and inhaled. EPA and FEMA have developed Protective Action Guides (PAGs) to indicate when long-term relocation of individuals should be considered. PAGs are primarily based on an assessment of the risk of developing cancer over an exposed individual's lifetime. They assume, conservatively,

that individuals are unsheltered and remain in the area during the entire period described for each contour. Contours show where individuals, if not relocated per the PAG, are projected to receive at least a specified dose in a specified time, as follows: inner contour (red), dose in first year post-attack, >2.00 rem; middle contour (orange), dose in second year post-attack, >0.500 rem; and outer contour (yellow), cumulative dose in the first 50 years post-attack. >5.00 rem. The cigar-shaped plumes often seen in models of atmospheric dispersion occur for gases or very fine particles, which would be the case for chemical warfare agents or fallout from a nuclear weapon but not in the case depicted. Whether such plumes would occur for an RDD depends on such factors as wind speed, type of explosive, and particle size.
(Provided by CRS): This note compares lifetime incidence of, and deaths from, cancer to those resulting from the attack modeled in this Figure. For the United States, the lifetime risk of being diagnosed with cancer is 43.61 percent, and the lifetime risk of dying from cancer is 21.15 percent. (U.S. National Institutes of Health. National Cancer Institute. Surveillance Epidemiology and End Results (SEER). "SEER Cancer Statistics Review 1975-2007," Tables 1.14 and 1.17, http://seer.cancer.gov/csr/1975_2007/results_merged/topic_lifetime_risk.pdf) For the 125,000 people in the affected area, the estimated lifetime incidence of cancer would thus be approximately 54,513 people, and the estimated lifetime deaths from cancer, 26,438. The attack would increase the lifetime incidence of cancer by 461 people, and lifetime deaths from cancer by 314. The Figure assumes no relocation, sheltering, or decontamination. All these actions would occur in the real world, significantly reducing cancer incidence and deaths caused by the attack.
This figure is from CRS Report R41890, *"Dirty Bombs": Technical Background, Attack Prevention and Response, Issues for Congress*, by Jonathan Medalia.

Figure 1. A Possible Radiological Dispersal Device (RDD) Attack on Washington, DC Using 1,000 Curies of Cesium-137 Chloride.

Cesium-137 is radioactive; cesium-137 chloride is of particular concern as a potential threat:

> Because of its dispersibility, solubility, penetrating radiation, source activity, and presence across the United States in facilities such as hospitals, blood banks, and universities, many of which are located in large population centers, radioactive cesium chloride is a greater concern than other Category 1 and 2 sources for some attack scenarios. This concern is exacerbated by the lack of an avenue for permanent disposal of high-activity cesium radiation sources, which can result in disused cesium sources sitting in licensees' storage facilities. As such, these sources pose unique risks.[7]

Figure 1 shows the area that an RDD attack using 1,000 curies (about 50 grams) of cesium-137 chloride could contaminate under the specified conditions. That amount fits into a vial 2-1/4 inches high by 1 inch in diameter. For comparison, 2,700 curies of cesium-137 is a category 1 quantity and 27 curies is a category 2 quantity.

REGULATORY BACKGROUND

The statute that created the Nuclear Regulatory Commission (NRC)[8] made it responsible for regulating the use and protection of special nuclear material—which for practical purposes consists of uranium enriched in isotope 235 (highly enriched uranium, or HEU, and low-enriched uranium, or LEU) and plutonium—in civilian uses, but not for regulating nuclear weapons or special nuclear material for weapons use. NRC also regulates fuel in nuclear reactors, which is typically low-enriched uranium, and "byproduct material," which includes many types of radioactive material other than plutonium or enriched uranium.[9] Byproduct material is widely used in medicine, industry, and research. NRC issues licenses permitting facilities (such as nuclear power plants or hospitals) or other organizations (such as industrial radiography companies that use mobile radioactive sources in the field) to possess specified quantities of radioactive material. A license is not needed to possess minute quantities of radioactive material, such as are found in home smoke detectors.

Agreement States are a central component of the nation's regulatory framework. Pursuant to the Atomic Energy Act of 1954, as amended (42 U.S.C. 2011 et seq.), NRC is responsible, with respect to radioactive material, for maintaining the health and safety of the public and the common defense and security. Section 274(b) of the Act permits NRC to relinquish its authority to regulate certain types of radioactive material within their borders to states that have entered into agreements with NRC for this purpose. There are 37 of these "Agreement States."[10] NRC can relinquish its regulating authority in the area of public health and safety; however, pursuant to Section 274(m) of the act, NRC cannot relinquish its regulatory authority in matters concerning the common defense and security to Agreement States. Each Agreement State regulates radioactive materials within its boundaries, which can include most medical, academic, and industrial uses of byproduct material. Agreement States do not regulate commercial nuclear power plants, research and test reactors, and facilities that are part of the nuclear fuel cycle.

In the wake of the 9/11 attacks, NRC issued orders to its licensees to enhance the security of radioactive materials. These orders went, among others, to licensees of large irradiators, manufacturers and distributors of radioactive material, and certain transporters of radioactive material, as detailed in **Appendix C**. An "Increased Controls" (IC) order went to other licensees authorized to possess specified quantities of certain radioactive material.

In the *Federal Register* of June 15, 2010, NRC solicited comments for a proposed rule, "Physical Protection of Byproduct Material," that would incorporate and modify previous security orders as 10 C.F.R. 37.[11] The proposed rule would deal with "the security requirements for use of category 1 and category 2 quantities of radioactive material."[12] NRC approved the rule in March 2012 and sent it to OMB for its review in July. NRC anticipates that the rule will be published in the *Federal Register* in early 2013.[13] For states regulated by NRC, the rule will become effective one year after it is published in the *Federal Register*. For Agreement States, the rule will take effect two years later. Once NRC and the Agreement States implement the rule, NRC and, when appropriate, Agreement States will rescind the various orders the rule replaces.

There are several differences between an order and a rule. An order can be effective immediately. The rulemaking process may take years, as it involves issuing a preliminary draft, soliciting comments, in some cases obtaining approval from the Office of Management and Budget, and a waiting period before the rule takes effect. An order that addresses an urgent situation, such as for post-9/11 security, may have little or no feedback from stakeholders, while a rule, such as 10 C.F.R. 37, entails extensive feedback, often with hearings, working groups, and lengthy correspondence. Orders apply only to current licensees; having an order apply to future licensees requires notifying them individually that the order applies to them. In contrast, a rule applies to current and future licensees. An order and a rule both have the effect of law. NRC wanted to replace its orders with a rule for various reasons:

> The orders issued by the NRC could stay in place indefinitely. However, the regulations would not reflect current Commission policy or requirements. Imposing long-term requirements through orders has not traditionally been the agency's preferred method of regulation. Orders, unlike rules, do not apply prospectively to applicants for new licenses. The NRC would have to periodically issue new orders to cover new and amended licenses. In order to make the requirements generally applicable to all present and future licensees, the security-related requirements need to be placed in the regulations. In addition, notice and comment rulemaking allows for public participation and is an open and transparent process.[14]

The rule makes many changes as compared to the orders, such as requiring individuals already deemed trustworthy and reliable (T&R) to complete certain security training, requiring a personal history disclosure, and requiring each licensee to "review the access program content and implementation" at least once a year. NRC provides details in a 69-page table comparing the rule and orders.[15]

NRC's policy is to publish an implementation guidance document in the *Federal Register* when, or shortly after, it issues a rule. Such documents provide details (often in the form of questions and answers) on how licensees are to implement a rule. The guidance, unlike a rule, is not mandatory, and licensees may find other ways to implement a rule, but guidance shows what NRC finds acceptable. NRC issued draft guidance in June 2010 for 10 C.F.R. 37;[16] however, there have been many changes to the rule since then, so there will be many changes to the final guidance document as well.

10 C.F.R. 37: Summary and Comments

10 C.F.R. 37 is included in an NRC document.[17] It is divided into six Subparts (with a seventh, Subpart E, reserved) and an Appendix. This part of the report comments on selected sections of Subpart A (General Provisions), B (Background Investigations and Access Control Program), C (Physical Protection Requirements During Use), and D (Physical Protection in Transit). Comments are based on interviews with radiation professionals. Two striking features of the interviews are the extent to which there is a diversity of views on almost every issue raised, and the many ways in which the rule impacts the radioactive-materials industry. Subparts F (Records) and G (Enforcement) and the Appendix (Category 1 and Category 2 Radioactive Materials) are brief and straightforward, so are not discussed here.

Subpart A: General Provisions

§ 37.1, Purpose: The purpose of the rule is to "provide the requirements for the physical protection program for any licensee" possessing certain quantities of certain radioactive material in order to "provide reasonable assurance of the security" of this material. While an RDD attack using less than a category 1 source of material could impose costs of billions of dollars in lost productivity and remediation, the requirements seek to provide only "reasonable assurance of the security" of such material. Issues are whether this standard is high enough and, conversely, what would be the cost vs. benefit of imposing a higher standard.

§37.5, Definitions: Two definitions are of particular interest. (1) The rule provides a complex technical definition of "byproduct material," which is what the rule protects. In brief, the term refers to various categories of radioactive material. Appendix A provides the complete definition.

(2) NRC states that "trustworthiness and reliability," or T&R, "are characteristics of an individual considered dependable in judgment, character, and performance, such that unescorted access to category 1 or category 2 quantities of radioactive material by that individual does not constitute an unreasonable risk to the public health and safety or security." Note that the T&R definition is subjective, and requires the person or persons making the T&R adjudication to decide if an individual is "considered" "dependable in judgment, character, and performance," and if unescorted access to certain radioactive material by that person "does not constitute an unreasonable risk to the public health and safety or security." Since it is up to each licensee to formulate its own criteria for adjudicating T&R, that approach inevitably results in non-uniform criteria, so that an individual might be granted T&R status by one licensee and not by another.

An issue is whether it is reasonable to expect persons without extensive training in deciding whether or not someone is a security risk, such as radiation safety officers (RSOs) [18] or human resources (HR) personnel, to make such determinations, and if such persons can do so reliably. Opinion on the part of radiation professionals varied widely on this point. An RSO at a large university said, "the idea that any university could determine if someone is a terrorist is silly. Universities are not set up for that." An RSO at another large university described a process in which an outside personnel security firm, for $125, runs background checks on individuals seeking T&R status. The firm gives its findings to university HR personnel, who say that they have no question about their ability to review T&R candidates because that is part of their job. The RSO then makes the final adjudication. An Agreement State director felt that licensees are best positioned to adjudicate T&R because they know their employees well, "so they can evaluate their character more easily and accurately than someone in [his state's capital] or Washington." He emphasized that the ability of small licensees to judge T&R is critical because a single misjudgment could cause major problems for the company, if not result in its end. Yet another university RSO felt, "having a federal agency make T&R adjudications would eliminate bias, as the adjudicators would not know the individuals being submitted for T&R."

§37.13, Information collection requirements: OMB approval: The rule imposes many such requirements. Some radiation professionals stated that 10 C.F.R. 37 placed a much more extensive burden of information collection on them than did the IC order; some felt otherwise.

An RSO at a large university:

> 10 CFR 37 is very administratively driven. There are a great many administrative requirements, such as for record-keeping, written security plans, and a requirement for annual reviews of these plans. Some of these are excessive, but we agree with other parts of 10 CFR 37. ... Regarding the security plan that the new rule requires (§37.43), we didn't write the plan under the orders because anything we documented would have to be

protected, and it's very difficult to impose document security on a university campus. A concern is that writing a detailed security plan creates a vulnerability: if a terrorist gained access to it, the plan could outline the strengths and potentially weaknesses.

An RSO at another large university:

NRC doesn't typically regulate by orders to licensees but decided to do so after re-evaluating the threat environment in the wake of 9/11, I feel the orders were the best that NRC could do at the time. However, I find 10 CFR 37 too burdensome. The rule is one size fits all, which causes a lot of problems. For example, radioactive sources in medical therapy units may be hard to steal, but have the same security requirements under the rule as self-shielded irradiators, which may be easier to break into. ...

Because of the difficulties of complying with the rule, I am urging researchers and users at my organizations to plan to switch to x-ray based irradiators rather than current ones that use radioactive materials. Since x-ray-based irradiators do not use radioactive material, they do not fall under the IC orders or 10 CFR 37. A blood bank I oversee plans to move to a new facility. It will replace one cobalt-60 irradiator with two x-ray irradiators; two are needed because of operational reliability issues that do not arise with radioactive material irradiators.

An Agreement State regulatory official:

I think that we are all working under what Past Commissioner McGaffigan stated was the mandate from the Commission regarding security: 'To provide reasonable assurance of adequate protection, not absolute assurance of perfect protection.' Under that premise, it is my belief that the burden or costs of implementing the previous orders or Part 37 are not too great.

Subpart B: Background Investigations and Access Authorization Program

§37.21, Personnel access authorization requirements for category 1 or category 2 quantities of radioactive material: Requires licensees with a certain amount of radioactive material to "establish, implement, and maintain its access authorization program." The objective of this program is to ensure that individuals with unescorted access to category 1 or 2 quantities of radioactive material and "reviewing officials" (personnel who make T&R adjudications) are T&R. There is a catch. Section 37.21(c)(3) states, "Licensees shall approve for unescorted access to category 1 or category 2 quantities of radioactive material only those individuals with job duties that require unescorted access to category 1 or category 2 quantities of radioactive material." Further, reviewing officials must be T&R. However, the Energy Policy Act of 2005 amended §149 of the Atomic Energy Act of 1954 to direct NRC to require that each individual who is permitted unescorted access to radioactive material as determined by NRC, or who is given access to "safeguards information," i.e., information such as on security plans for protecting radioactive material, be fingerprinted, with the fingerprints submitted to the Attorney General (in effect, the FBI) for a criminal history record check. Often, however, T&R adjudications are done by human resources personnel, who do not need unescorted access to radioactive material or access to safeguards information to do their job. Yet since the law requires fingerprinting and a criminal history record check only of individuals requiring such access, some organizations have given HR staff unescorted access in order to enable them to be fingerprinted. One university solved this dilemma by granting HR staff unescorted access but not giving them the requisite key card that would give them physical access! One individual expressed a need for NRC to seek to have the law changed to avoid this dilemma. Another university RSO saw an imbalance between T&R and physical security:

> In general, I believe that limited funds should be spent more to augment physical security, especially internally to the devices, and less on T&R. Indeed, by spending more on physical security, there is less concern about T&R. If devices are made harder to penetrate, or if there are more alarms, it would be harder for anyone to steal radioactive material.

NRC staff elaborated on the link between fingerprinting and unescorted access:

> Section 652 of the Energy Policy Act of 2005 (EPAct) amended Section 149 of the Atomic Energy Act of 1954, as amended (AEA). The AEA requires the NRC to fingerprint any individual permitted unescorted access to (1) a utilization facility; (2) radioactive material or property regulated by the Commission that the Commission has determined to be of such significance to the public health and safety or common defense and security as to warrant fingerprinting and background checks; or (3) is permitted access to safeguards information as defined in Section 147 of the Atomic Energy Act of 1954, as amended. *The NRC does not have authority to fingerprint any other individuals.*
>
> The NRC has determined that it needs to have confidence in the integrity of the reviewing official because that official approves individuals for unescorted access category 1 and category 2 quantities of radioactive materials. Accordingly, the NRC requires that the individual be fingerprinted and undergo a background investigation and be determined to be trustworthy and reliable. However, as noted above, pursuant to Section 149 of the AEA, the NRC may only fingerprint those individuals who are permitted unescorted access to a utilization facility, certain radioactive materials or property regulated by the Commission, or safeguards information. This is why an individual designated by the licensee to serve as a reviewing official must be permitted unescorted access to category 1 and category 2 quantities of radioactive materials or to safeguards information.
>
> The reviewing official is one of the layers for defense-in-depth of the security program. If the reviewing official is not fingerprinted, a gap could be created in the security program that could potentially be exploited. The reviewing official could have a criminal history or terrorist ties and allow other individuals with a criminal history or terrorist ties to have unescorted access to radioactive material in quantities of concern.[19]

§37.23, Access authorization program requirements: (1) States that "reviewing officials" are the only individuals who may make T&R determinations, and requires the licensee to "provide under oath or affirmation, a certification that the reviewing official is deemed trustworthy and reliable by the licensee." Not stated is how the licensee is to make this determination, and what assurance there is that licensee personnel making this determination are qualified to do so.

(2) Requires licensees to "develop, implement, and maintain written procedures for implementing the access authorization program." Each licensee is to develop its own procedures, which inevitably means that the procedures will differ from one licensee to another. Several interviewees expressed concern about this point. For example, if an individual moves from one organization to another, the first organization can transfer to the

second the information on the individual's background investigation but cannot transfer its adjudication of the individual as T&R. Several interviewees stated that NRC has not developed a uniform set of standards or guidelines for adjudicating someone as T&R, and found this to be quite frustrating. A representative of the Nuclear Energy Institute said, "NRC does not want to adjudicate T&R even for difficult cases, and it has not even put out criteria that would disqualify someone as T&R even though that would be helpful. At a minimum, NRC needs to provide clear guidance on T&R to its licensees." An official at a company that manufactures and distributes radioactive sources echoed similar sentiments, and noted that the T&R requirement imposes real burdens, such as requiring the company to actively seek out references for T&R applicants who were not listed in the applicant's paperwork. This official stated, "NRC did not even provide guidelines on what factors should lead to rejecting someone as T&R."

A statement by an Agreement State regulator may help resolve the discrepancy between those saying that T&R adjudications are best done by employers but that NRC has not provided adequate guidelines for such adjudications: "The licensees begged NRC to provide T&R criteria. They asked over and over again, including for examples, but NRC wouldn't do it, though it's not clear as to why. At this point, however, licensees have so much experience in adjudicating T&R that there would not be much value added if NRC provided criteria."

§37.25, Background investigations: This section sets minimum requirements for a background investigation, including fingerprinting and an FBI identification and criminal history records check; verification of true identity, employment history, and education; and a character and reputation determination. Regarding the latter point, "reference checks under this subpart must be limited to whether the individual has been and continues to be trustworthy and reliable," yet it is unclear whether the references are able to judge whether the applicant is T&R. An issue is whether an applicant undergoing a background investigation would give as references individuals who would not vouch for him or her; accordingly, §37.25(a)(6) states, "The licensee shall also, to the extent possible, obtain independent information to corroborate that provided by the individual (*e.g.*, seek references not supplied by the individual)." Yet one interviewee said it was very difficult to find references not listed by the applicant. A draft of the rule had a requirement for a credit check, but many stakeholders opposed that provision on grounds that a person's credit history—especially in the current economy—has no bearing on whether he is T&R. For example, a person could have lost her job, or gone bankrupt because of medical bills; such events would not indicate lack of T&R. NRC dropped that provision from the final rule.

§37.27, Requirements for criminal history records checks ...: This section requires licensees seeking to grant unescorted access to category 1 or 2 quantities of radioactive material to individuals to fingerprint the individuals and transmit the fingerprints to NRC. NRC will then transmit the fingerprints to the FBI, which will conduct a criminal history records check. The FBI then transmits the criminal history records check to NRC, which in turn transmits it to the licensee. The licensee uses this information as part of the required background investigation to determine whether or not to grant unescorted access to the individual. Radiation professionals felt that this procedure is cumbersome; they should be able to send fingerprints directly to the FBI, and receive information directly from the FBI. Two interviewees complained that they wanted to collect fingerprints electronically rather than by using fingerprint cards, but said that they could not do so because of the rule. While §37.27(c)(1) specifically permits use of an electronic fingerprint scan, a university RSO said

> my institutions can only use NRC fingerprint cards, at a cost of $26 per FBI background check request (if more than two cards are needed by FBI, up to an additional $52 may be required). When establishing the required fingerprinting program, I investigated use of a vendor to provide electronic fingerprinting service similar to that used by the state's Department of Elementary & Secondary Education. I ultimately learned that the NRC would only accept electronic fingerprints if sent directly from the licensee, and not a vendor (equipment like this is cost prohibitive for materials licensees). I have had several hundred people go through FBI fingerprint background checks, which is quite burdensome. Why pay NRC to be the middleman instead of going directly to the FBI? Why can't one central federal agency make these kinds of T&R adjudications?
>
> NRC provided the following comment:
>
> 42 USC 2169 prescribes this process:
>
> (2) All fingerprints obtained by an individual or entity as required in paragraph (1) shall be submitted to the Attorney General of the United States through the Commission for identification and a criminal history records check.
>
> (3) The costs of an identification or records check under paragraph (2) shall be paid by the individual or entity required to conduct the fingerprinting under paragraph (1)(A).
>
> (4) Notwithstanding any other provision of law–
>
> (A) the Attorney General may provide any result of an identification or records check under paragraph (2) to the Commission; and
>
> (B) the Commission, in accordance with regulations prescribed under this section, may provide the results to the individual or entity required to conduct the fingerprinting under paragraph (1)(A).
>
> The law also permits electronic fingerprinting as follows:
>
> d. The Commission may require a person or individual to conduct fingerprinting under subsection a.(1) by authorizing or requiring the use of any alternative biometric method for identification that has been approved by–
>
> (1) the Attorney General; and
>
> (2) the Commission, by regulation.
>
> However the Department of Justice has determined that criminal history record information cannot be provided to any party other than the NRC, the licensee, or the individual that is the subject of the record. That is why the use of a third party contractor is an issue.[20]

§37.29, Relief from fingerprinting, identification, and criminal history records checks ...: This section provides that individuals in 13 categories do not need to undergo these checks in order to have unescorted access. Categories include Members of Congress; employees of NRC, of the executive branch, and of Members and committees of Congress who have undergone fingerprinting for a prior U.S. Government criminal history records check; federal, state, and local law enforcement personnel; state governors or their representatives; and others.

§37.31, Protection of information: This section requires licensees to protect personal information obtained through background investigations. It requires licensees to "establish and maintain a system of files and written procedures for protection of the record and the personal information from unauthorized disclosure." It also permits personal information to be provided to another licensee under certain conditions. While none of the interviewees expressed concern about protecting personal information, quite a few expressed concern about requirements throughout 10 C.F.R. 37 for creating and protecting written procedures. This issue is dealt with in more detail below, but this section is an example of what many interviewees saw as an administrative burden. Some interviewees also felt that it should be possible to transfer a T&R adjudication in addition to personal information. One said, "It also seems arbitrary that an employee moving to another company can have his or her background investigation data transferred to the new employer, but not the first company's adjudication of the person as T&R."

§37.33, Access authorization program review: Requires each licensee to "be responsible for the continuing effectiveness of the access authorization program. ... The review program shall evaluate all program performance objectives and requirements." An issue is whether each licensee has the ability to review its own program, and whether it can do so objectively. This section also requires licensees to document results of the reviews and any recommendations. Further, "Each review report must identify conditions that are adverse to the proper performance of the access authorization program, the cause of the condition(s), and, when appropriate, recommend corrective actions, and corrective actions taken." This is an administrative burden imposed by the rule.

Subpart C: Physical Protection Requirements during Use

§37.41, Security program: "Each licensee shall establish, implement, and maintain a security program that is designed to monitor and, without delay, detect, assess, and respond to an actual or attempted unauthorized access to category 1 or category 2 quantities of radioactive material." Specific requirements are detailed, and commented on, in following sections.

§37.43, General security program requirements: This section requires licensees to have a security plan, implementing procedures, training, and methods to protect information. The security plan shall be written and specific to each licensee's facilities and operations. The plan shall describe measures, and identify resources, for implementation. It shall be "reviewed and approved by the individual with overall responsibility for the security program." This individual shall also approve, in writing, the implementing procedures. In an effort to ensure that this individual and others have the training and expertise necessary to determine the adequacy of the security program, this section requires "each licensee [to] conduct training to ensure that those individuals implementing the security program possess and maintain the knowledge, skills, and abilities to carry out their assigned duties and responsibilities effectively." Training must, at a minimum, include instruction in the licensee's security program, the responsibility to report to licensee conditions that may violate NRC requirements, the responsibility of the licensee to report promptly to local law enforcement actual or potential threats to category 1 or 2 material, and how to respond to security alarms. Finally, this section requires licensees to protect security-related information, such as the security plan and implementing procedures, and requires "written policies and procedures for controlling access to, and for proper handling and protection against unauthorized disclosure of, the security plan and implementing procedures."

Interviewees offered many comments on these requirements. An RSO at a university stated,

> An NRC manager said NRC thinks that the new rule will not be much of a burden to licensees because they have already implemented many of the requirements in 10 CFR 37. I strongly disagree. The rule is one size fits all: security plans, training, procedures, new concepts (like security zones). There were orders, and we implemented them. After we had been inspected a few times, we were able to understand how to comply with the orders. Now with the new rule, we will have to set the rule as the new baseline. We will be reinspected against the new standards. It will take significantly more time. There are 37 Agreement States plus NRC, so there will be 38 interpretations of what the rule means. We will have to start all over again. A big problem will be the paperwork.

Regarding the requirement for a written security plan and written implementing procedures, the RSO continued,

> we didn't write the plan under the orders because anything we documented would have to be protected, and it's very difficult to impose document security on a university campus. A concern is that writing a detailed security plan creates a vulnerability: if a terrorist gained access to it, the plan could outline the strengths and potentially weaknesses. It is easier for NRC to inspect a written plan than to sit down and discuss (orally) what the plan is. This puts the burden on the licensee.

An RSO at another university said,

> there is a new requirement to train people every year on security. This is a large time commitment for individuals with unescorted or need-to-know access and for me as compared to the orders. For example, staff can have refresher courses in health and safety online, but security training cannot be done online because of the security nature of the material. Since the institutions for which I am the RSO have several hundred people currently granted unescorted access to category 1 and 2 material, several training sessions must be done so as to cover all the people needing training every year.

Regarding the requirement to protect information, an industrial radiographer said,

> Yet another concern I have is that 10 CFR 37 requires limiting information on security to people with a need to know, but in practice information is often widely disseminated. At the steel mill, all 200 electronics technicians know that there is a storage room where spare devices are stored. Same story in a hospital, where most of the doctors and nurses know if their hospital has a gamma knife. Limiting information can be tough to implement in practice.

An official at a manufacturing and distribution company noted a disparity in regulations protecting radioactive and other material:

> The nuclear industry is very heavily regulated, while other industries that pose significant health and safety risks are much less stringently regulated. For example, tanker trucks carrying large quantities of dangerous chemicals can go through large cities.

§37.45, LLEA [local law enforcement agency] coordination: Licensees "shall coordinate, to the extent practicable, with an LLEA for responding to threats to the licensee's facility, including any necessary armed response." Among other things, the licensee must notify LLEA "that the licensee will request a timely armed response by the LLEA to any actual or attempted theft, sabotage, or diversion of category 1 or category 2 quantities of material." The licensee shall notify NRC if LLEA has not responded to the request for coordination within 60 days, or if LLEA does not plan to participate in coordination.

Many interviewees reported excellent cooperation between licensees and LLEA. An Agreement State official found that the security orders (preceding 10 C.F.R. 37) promoted working relationships:

> before NRC issued its security orders, there was often not a good relationship, if any, between licensees and LLEA. The orders, which require licensees to contact LLEA to try to work out a response plan, have in many cases led to the establishment of relationships between licensees and LLEA.

Initially, he said,

> regulators and licensees came to the orders with limited security real world/operational experience, but believe training, knowledge gained and partnerships formed while under the orders prepare everyone well for the implementation of 10 CFR 37. The orders and the new rule are designed to motivate the integration of LLEA, by both regulators and licensees. LLEAs bring that real world security experience to the final safety culture product.

He found LLEAs receiving homeland security grants provided by the U.S. Department of Homeland Security are more sensitized to potential threats from radioactive materials, and said, "LLEA is absolutely much more aware of the need for radioactive material security post-9/11."

A second Agreement State official found cooperation improving:

> In the beginning, LLEA was not so cooperative in coordinating security plans with licensees; this was a common violation. Often, they were not interested. But as LLEA awareness of radioactive threats increased and as regulators tried to help licensees contact the right people in law enforcement, LLEA now "gets it" and wants to know if there is radioactive material in their jurisdictions.

A third Agreement State official noted that LLEA work closely with licensees, conduct security inspections[21] to facilities housing radioactive material, have alarms linked to police headquarters, and provide armed escort of radioactive material within the state's borders. Some police departments, including those in some large cities, send all their officers to Advanced Radiological Response Techniques training courses at the Y-12 National Security Complex in Tennessee. LLEA in states such as these would give high priority to a theft of radioactive material.

On the other hand, other interviewees reported instances of noncooperation. In one case, LLEA rebuffed repeated attempts to contact it; the interviewee found the situation "unworkable." Another said that LLEA was unaware of the issue of radioactive material security. Further, circumstances of budget and distance may conspire to render LLEA less cooperative than they might like to be. An Agreement State official noted this in his state, and pointed out that there is no mechanism to force LLEA to cooperate, let alone to the extent NRC wants:

> Local law enforcement agencies (LLEAs) are cooperative up to a limit. They want to cooperate but don't want to be pinned down to certain requirements, such as responding within [a specified time]. NRC would like the theft of a radioactive source to be high on the LLEA priority list, but that probably won't happen. Some state troopers cover several hundred miles on their routes. If an event happens at one end of their route, it's unrealistic to expect them to drop everything and rush to the other end in [a specified time]. LLEA is committed to assisting if feasible, but if there's a bank robbery or other emergency situation, should they disregard it to respond to a radiological event? They need to make a choice. As a regulator, neither the NRC nor the Agreement States have any control over LLEA. Licensees also have no control. They try to get LLEA to participate, and document their efforts to do so.

§37.47, Security zones: Licensees shall ensure that category 1 and 2 materials are within security zones established by the licensee. Security zones shall allow unescorted access to approved individuals in several ways: (1) "Isolation of category 1 and category 2 quantities of radioactive materials by the use of continuous physical barriers that allow access to the security zone only through established access control points. A physical barrier is a natural or man-made structure or formation sufficient for the isolation of the category 1 or category 2 quantities of radioactive material within a security zone." (2) By direct control of the security zone, or (3) by both methods.

There are potential issues with the isolation approach. First, the definition is tautological: isolation is to be achieved by use of barriers sufficient for isolation. Second, there is no requirement for the amount of control, if any, that must be present at an access control point. In particular, would it be vulnerable to an armed attack by terrorists?

An industrial radiographer said,

> 10 CFR 37 will be burdensome in terms of cost and what it requires. I am particularly concerned about security zones. Most of my clients are in industries, such as steel and petrochemicals, and use radioactive sources in the form of gauges, such as to measure thickness of steel or flow of chemicals.

He noted that a steel mill inherently has many features that would make it exceedingly difficult for terrorists to steal radioactive sources, and security devices such as alarms and cameras.

> The security features already in place should be sufficient to keep people from gaining unauthorized access to the material. It seems excessive to require the licensee to set up a security zone, which is a barrier that absolutely prevents people from getting to that area. It is even more burdensome for licensees if the security zone extends well beyond the gauge itself.

An Agreement State official said, "The rule establishes security zones, which are not in the [Increased Controls order] and are confusing to the Agreement States as to where a zone must be established. Licensees will have to learn about establishing these zones."

On the other hand, another Agreement State official is pleased that the rule is not overly prescriptive, i.e., that it does not specify in minute detail what must be done. As a result, the rule "leaves us a lot of room to work. For example, not telling me how to do security zones is fine."

§37.49, Monitoring, detection, and assessment: "Licensees shall establish and maintain the capability to continuously monitor and detect without delay all unauthorized entries into its security zones." In contrast to much of the rest of 10 C.F.R. 37, this section is prescriptive and detailed. Examples of methods for monitoring and detection include a monitored intrusion detection system linked to a central monitoring facility, and a monitored video surveillance system. Licensees must also be able to detect unauthorized removal of radioactive material from the security zone. For category 1 material, this can be achieved by electronic sensors linked to an alarm or direct visual surveillance. For category 2 material, it can be achieved by weekly verification through physical checks. Specified other means for monitoring and detection may also be used. The section also requires immediate assessment of actual or attempted unauthorized entry; maintenance of communication and electronic data transmission, including an alternative communication capability; and requesting an armed response by LLEA involving theft, sabotage, or diversion of category 1 or 2 materials.

Some interviewees were unsure how many, and what types of, security features they needed to install in order to satisfy these requirements. Others consulted security professionals who designed security systems for their facilities. Many felt that NNSA's Global Threat Reduction Initiative, described below, was important in helping them set up a security system.

An industrial radiographer expressed concern about the requirement for a weekly inventory of category 2 sources:

> the rule takes us from a quarterly inventory of sources to a weekly inventory. On the surface, this is not a bad idea. For me, this is not a problem because most of my sources come back each night, and there is a benefit of source security, which is a top concern for us. But [another] company ... works in all 50 states and its sources do not return to a centralized location. This provision of Part 37 is a nightmare for them, imposing a huge logistical burden.

The requirement for weekly verification of category 2 sources to ensure that the material is present is found in 10 C.F.R. 37.49(a)(3)(ii). NRC notes that the Increased Controls order does not require a quarterly inventory of category 2 sources.[22]

An official with a company that manufactures and distributes radioactive sources expressed a similar concern:

> One example of the burden is a requirement to do a weekly check of all sealed radioactive sources. This may be workable for a company with a few dozen sources, but [our company] has thousands of sources. It would be impossible to check them all each week, and someone doing the checking would receive too much radiation exposure.

A university RSO favors the British model for deciding on the adequacy of physical protection measures:

> In discussions with colleagues from Great Britain, all hospital security plans are centrally reviewed by a police agency PRIOR to paying for the new equipment or installations. Therefore, there is agreement between groups of the necessary security before any money is spent. While not every site is the same and some flexibility is required, it would be really nice to know if the systems being paid for will be deemed adequate before they are installed.

NRC notes that the U.S. procedure is different. Rules are printed in the *Federal Register.* NRC issues a guidance document to accompany each rule, and prints a notice of availability of this document in the *Federal Register.* The rule and notice are often, but not always, printed in the *Federal Register* at the same time. The guidance document, while not mandatory, provides detailed questions and answers on what constitutes compliance. The appropriate regulatory authority—NRC or an Agreement State—does not approve the licensee's security measures in advance. Once the measures are in place, regulators inspect the facility to determine whether they meet the requirements of the rule. Since the guidance is not mandatory, licensees may adopt other approaches that they believe to be acceptable ways of complying with the rule. The authority would then inspect the facility once the measures have been implemented to ensure that the approach, as implemented, complies with the rule.[23]

§37.51, Maintenance and testing: Licensees shall maintain physical components of the security systems, and shall test them at the frequency the manufacturer suggests. If there is no such suggestion, the frequency shall be at least once every 12 months. The RSO at a large university found this requirement problematic:

> An earlier version of Part 37 required quarterly testing of alarms, or as recommended by the manufacturer. Quarterly was too frequent because of the number of alarms and the amount of documentation. Some items can break or be knocked out of alignment if tested, such as magnetic door position switches. And some plastic access controls can be damaged during disassembly to test the internal tamper alarm. These really do not have any method for breakage so they don't really need to be tested. Furthermore, testing can be complicated, as it sends alarms to police, distracting them from their work. Section 37.51 of the final rule calls for testing in accordance with the manufacturer's recommendation, and if no recommendation then at least every 12 months. One manufacturer states: "This security system requires very little maintenance, however, test the system weekly to ensure it is working properly" The company then goes on to offer to sell the testing service to the customer. ... Thus, the change in the final rule is not an improvement. In some circumstances it requires much more frequent testing than the Draft Part 37 required.

The RSO suggested an alternate wording: "The equipment relied on to meet the security requirements of this part must be inspected and tested for operability and performance at least annually."

§37.53, Requirements for mobile devices: Licensees with mobile devices containing category 1 or 2 quantities of radioactive material must "have two independent physical controls that form tangible barriers to secure the material from unauthorized removal." None of the interviewees expressed concern about this provision. Security of mobile sources has not been a problem. An industrial radiographer said,

> Up until 2011, no one had stolen a source, though people occasionally stole trucks that incidentally had small radioactive sources on them. Last year, a radiography camera was stolen in Texas and was never recovered. However, since its radioactive material was iridium-192, with a half-life of 74 days, over 99 percent of that radionuclide will have decayed in about two years. That source, 33 curies of Iridium 192, would be approximately 2 curies now, so this clearly was not a theft for terrorist purposes.

§37.55, Security program review: Licensees shall ensure that the security program is reviewed, at least annually, with the results and recommendations documented, and shall take actions to correct adverse conditions. An RSO at a large university found some administrative requirements to be excessive, but an RSO at another large university found the burden worth the effort:

> At first glance, 10 CFR 37 appeared to impose additional burdens compared to the Increased Controls Order. For example, it required security system reviews each quarter and a performance review every year. Despite the burden, ... the university police department mandated these periodic security system reviews two years ago. The university police has experts in security, security systems and a capable police force, so their judgment on security matters is well-respected in the local law enforcement authority (LLEA) community. The security reviews have proven to be of great value.

§37.57, Reporting of events: §37.57(a) requires licensees to immediately notify LLEA of actual or attempted theft, sabotage, or diversion of category 1 or 2 material, and also to notify NRC within four hours of discovering such event. §37.57(b) requires licensees to assess suspicious activity of this nature, notify LLEA as appropriate, and also notify NRC within four hours of notifying LLEA.

There was one comment on this latter provision. An RSO at a large university had an experience in which these requirements were excessive:

> An unknown State Police Officer in uniform in his private car drove up to our reactor and requested admittance (wearing a side arm). This was very suspicious. The LLEA was notified. The LLEA sergeant responded very quickly. He knew the individual, a State Police captain. The State Police captain promised to NEVER do that again. According to §37.57 of the final rule, I must still notify the NRC within four hours.

The RSO feels that the rule should offer some flexibility as to what incidents to report. If an event occurs that initially appears suspicious but proves not to be, there should be no requirement to contact NRC. NRC comments that there is no requirement to notify NRC if, after assessing the situation, the licensee determines that there is no need to contact LLEA. The licensee only needs to contact NRC about suspicious activities if it has contacted LLEA.[24]

Subpart D: Physical Protection in Transit

Comments in this section are mainly from a shipper of radioactive materials. A key assertion he makes, not reflected in a section-by-section analysis of Subpart D, is that Subpart D should have been placed in Title 49 of the U.S. Code (transportation) rather than in Title 10 for the following reasons:

> the Agreement States will have some flexibility in how they implement the Part 37 requirements. This flexibility may result in Agreement State implementation of Part 37 Subpart D regulations in a manner that would otherwise result in preemption if the regulations were contained in Title 49. For example in the early 1990's the State of Washington attempted to establish highway routing restrictions for radioactive materials that limited the locations that trucks hauling radioactive material could enter the State. This regulation was preempted by the DOT under the dual compliance and obstacle tests, reference 58 FR 31580. My concern is that given the pre-planning and coordination requirements contained in Part 37 Subpart D, an Agreement State could effectively restrict the routing of radioactive materials through the State. Another concern is that states may charge unreasonable fees for transporting category 1 and category 2 materials through the state, in these cases the NRC cannot challenge the fees, but the DOT could. Differences in escorting and vehicle inspections requirements could also place undue burdens on the shippers and carriers of category 1 and 2 quantities of materials. Generally speaking, when transportation requirements are set by many regulators and by federal interstate commerce rules as well, compliance with transportation regulations can become very complicated.

§37.71, Additional requirements for transfer of category 1 and category 2 quantities of radioactive material: Licensees transferring such material shall verify that the recipient's license authorizes receipt of the material. "In an emergency where the licensee cannot reach the license issuing authority and the license verification system is nonfunctional, the licensee may accept a written certification by the transferee that it is authorized by license to receive the type, form, and quantity of radioactive material to be transferred. ... The certification must be confirmed by use of the NRC's license verification system or by contacting the license issuing authority by the end of the next business day." A shipper found it generally straightforward to verify the recipient's license, and noted that, in his experience, the emergency provision did not come into play:

> §37.71(c) provides an alternative method for obtaining permission in an emergency situation, but I don't know when that would arise. We haven't had that happen. We typically plan category 1 shipments months in advance; sufficient advanced planning is associated with category 2 shipments as well.

§37.73, Applicability of physical protection of category 1 and category 2 quantities of radioactive material during transit: This section refers licensees to at least 44 provisions of 10 USC 37 and 71. Some apply to imports and exports, others apply to category 1 or category 2.

A potential issue is the degree of complexity of complying with the many provisions referenced in this section. A shipper felt that it was not difficult to comply with the provisions of Subpart D because his company had to implement many of these provisions as a result of previous NRC orders, notably for radioactive material quantities of concern (RAMQC).[25] He said,

> Subpart D does not add burdens beyond the existing NRC orders. There are minor things, such as updating the transportation security plan, and developing procedures to ensure compliance with the regulation but these are administrative activities that we do not consider burdensome because the Part 37 requirements for route notifications, estimation of when a shipment is going to cross state borders, delivery time, revised delivery time if there are changes, and so on are already in the RAMQC order.

§37.75, Preplanning and coordination of shipment of category 1 or category 2 quantities of radioactive material: Licensees planning to transport, or to have transported, category 1 material shall conduct detailed preplanning and coordination of the shipment with the recipient, and any state through which the shipment will pass, identify safe havens along the planned route, and document these activities. Requirements for category 2 shipments are less stringent. As noted under §37.73, the shipper did not consider these requirements burdensome because they are already in the RAMQC order. According to NRC, "A safe haven is a readily recognizable and readily accessible site at which security is present or from which, in the event of an emergency, the transport crew can notify and wait for the LLEA. The NRC expects safe havens to be identified and designated by the licensee."[26] However, the shipper noted that NRC "provides for safe havens, but doesn't provide a list of safe havens. It is up to us to identify safe havens."

§37.77, Advance notification of shipment of category 1 quantities of radioactive material: Licensees planning to ship category 1 material shall provide advance notification of the shipment to NRC and the governor, or the governor's representative, of each state through which the shipment will pass, using procedures this section details.

§37.79, Requirements for physical protection of category 1 and category 2 quantities of radioactive material during shipment: This section spells out requirements for shipment by road and rail in great detail. For example, a licensee transporting category 1 quantities of material by road shall ensure that movement control centers and redundant communications are established, and that shipments are "continuously and actively monitored by a telemetric position monitoring system or an alternative tracking system reporting to a movement control center." For shipments of category 1 quantities of material by rail, the licensee shall "ensure that rail shipments are monitored by a telemetric position monitoring system or an alternative tracking system."

While shipment by truck entails a driver in close proximity to the radioactive material, that is not the case for rail shipments. In theory, this might constitute a vulnerability. In practice, however, the shipper said that this section "provides for security of transportation of category 1 and 2 radioactive material by rail, but I don't know of any instances where such material is transported by rail." The shipper also noted that §37.79 does not address shipments by air for imports or exports of radioactive material. "Part 37 applies to the domestic portion of the highway (or rail) shipment, so once we deliver the shipment to the airport we are relieved of the transportation security burden." In contrast,

> Imports and exports by sea are much more difficult because we have to coordinate the arrival or departure of our shipment with the arrival or departure of a ship. This is difficult because ships may encounter delays enroute. We planned an export to South

> Africa, to be carried on a ship from Baltimore. The ship was delayed for 48 hours. The port would not accept the material, so the truck, with its two drivers, had to wait at a truck stop. If we could have delivered the material to the port, we would have been in compliance with the NRC order, and the port would have been much more secure than the truck stop.

§37.81, Reporting of events: The licensee doing the shipping shall notify LLEA and NRC within 1 hour of determining that a shipment of category 1 material is lost or missing, and shall notify NRC within 4 hours if category 2 material is lost or missing, as detailed in this section.

Additional Comments

Discussions with radiation professionals yielded comments on several topics that are not directly a part of 10 C.F.R. 37 but are related to it. They are discussed here because they add insight into how radiation professionals view, and are affected by, the new rule. This section presents views— often, contending views—on each of several topics from various radiation professionals.

The Rulemaking Process

The transition from security orders to a rule: Opinions differed on whether the orders should have been left in place, or should have been codified into a rule and subsequently modified, or combined into a new rule. A university RSO said:

> NRC says it is incorporating lessons learned in the rule, and that it issued orders because it had to get them out quickly. But if the orders provide adequate security, why not codify the orders and modify them later as needed? NRC is making fundamental changes to the orders in 10 CFR 37, with additional requirements and burdens for licensees. If the Orders were adequate, why change?

A Nuclear Energy Institute representative said,

> NEI's view is that NRC should have codified the security orders, and then done a careful analysis of enhancements beyond the orders, including vulnerability assessments.

An industrial radiographer said,

> To comply with the new rule, plans will have to be changed, and people covered by security orders will have to be retrained to the new terms. This will impose a financial burden on licensees, but without benefit. One of my clients is pulling his hair out because of the added burden. His feeling is that his organization has a good program, it has been inspected three or four times, and everything is up to snuff. Why change it? NRC should have simply codified the orders and then changed them as necessary.

According to an Agreement State official,

> Initially, most Agreement States would have preferred to simply codify the security orders. The new rule, however, is a consensus of Agreement State and USNRC ideas learned from experience while using the orders. A rule established and enhanced from experience would seem to have a better foundation than orders issued to address an immediate security need. The new rule also combines all security rules into one regulation. It is more efficient for regulators and the regulated community.

A university RSO favored a new rule, but finds it essentially a codification of the orders. He

> likes the fact that 10 CFR 37 combines multiple orders into one new part of 10 CFR rather than codifying the orders as they were. Having them all in one place makes it easier to follow than if a licensee had to look in multiple orders, revisions to orders, Q&A, revisions to Q&A, etc. In effect, by combining the orders and adding material, NRC codified them.

An Agreement State official pointed to another reason to have a rule rather than orders:

> My legal staff is of the opinion that it is easier to conduct enforcement of a rule than of an order. Specifically, the staff says, if a licensee violates a rule, the state is in a stronger position in a court case than if the licensee violates an order.

How well did NRC conduct the rulemaking process? Radiation professionals had mixed views on this point. Some found NRC to be responsive, others did not. An industrial radiographer:

> When [the Increased Controls order] was being crafted, it included a lot of provisions that wound up being jettisoned because of strong opposition. NRC included many of those items into a draft of Part 37 and, not surprisingly, they met with strong opposition. I and others testified before NRC on these points, and NRC accepted much of what we told them. But most of the industry didn't know about the draft Part 37 until very late. NRC could have sent notices to all its licensees, or to the Agreement States. I didn't see a strong NRC effort to pull in the stakeholders. ... I give kudos to the NRC for listening to the users and adjusting the regulation in response to our comments. I strongly believe that input from the "experts" in the field, i.e., the end users, should have a large part in regulatory affairs as it is those in the field that see the real impact of regulations.

A Nuclear Energy Institute representative:

> The current security orders include security requirements based on vulnerability assessments. NRC says it used lessons learned from implementing the security orders, but concedes that it had not done vulnerability assessments in preparing the rule. Without the assessments, the justification for the enhancements and inclusion of some categories of licensees is unclear. ... NRC issued security orders for panoramic and underwater irradiator licensees and for manufacturer and distributor licensees, and increased controls for other licensees. Each order provided separate requirements for each type of licensee. Licensees have been operating within the security orders. NRC merged the three sets of requirements and imposed this merged set on every licensee, including nuclear power

plants and licensees operating other parts of the nuclear fuel cycle. As a result, the rule is quite complicated. It would have been better if NRC had codified the orders and made revisions later in a risk-informed manner, as necessary, based on objective vulnerability assessments.

An official of the Conference of Radiation Control Program Directors:

CRCPD had serious concerns on the draft rule but the concerns have been removed. For example, NRC removed the requirement to check credit history as part of T&R; CRCPD felt that credit checks do not give good information and that other tools were sufficient.

An Agreement State official:

The state members of the rulemaking working group kept asking NRC if there was some particular threat, and if so, shouldn't they tailor the rule to the threat and the users. NRC did not provide an answer. As a result, the rule has overkill in terms of the level of security.

Another Agreement State official:

[Agreement State representatives] worked closely with NRC to develop the rule. We were co-regulators and had a good partnership with NRC. For example, we were involved in developing parts of the rule concerning fingerprinting and reviewing officials. We feel that the [Increased Controls order] and the rule provide reasonable assurance— not certainty— about the security of radioactive material. We are quite satisfied with the rule, and were able to get our input in; it was an excellent experiment in rulemaking. We feel that the rule is pretty clear.

NRC offered rebuttals to several of the foregoing comments. They are assembled here:

(1) The ICs themselves were not revised. NRC used stakeholder comments to develop the guidance document accompanying the ICs.

(2) The NRC conducted a number of outreach activities for the rule. First, the technical basis for subpart D was placed in the Federal Register for public comment. Three public meetings were held to receive comments. Second, the preliminary draft language for subparts B, C, and D were placed in the Federal Register for comment. Third, the draft final rule and guidance were placed in the Federal Register for comment. The comment period lasted for more than six months. By letter dated June 15, 2010, the NRC sent notification to every licensee that received an NRC order and everyone that commented on the preliminary rule language that the proposed rule was available for public comment (and enclosed a copy of the Federal Register notice that contained the proposed rule); the notice also provided information on the guidance document and the public meetings (2) that were planned on the guidance. The NRC also provided a copy of the letter to the Agreement States for their use and information. Many of the Agreement States provided the information to their licensees.

(3) While NRC did not perform a new vulnerability assessment in preparing the rule, the original vulnerability assessment provided security enhancements that are applicable to both the ICs and Part 37.

(4) All Orders were essentially the same. They all contained requirements to 1) control access, conduct trustworthy and reliable determinations, 2) detect, assess, respond to unauthorized access, and 3) transportation security requirements. The performance based

nature of the Orders and 10 CFR Part 37 allows licensees to tailor their security plan to their specific situation. The main difference is the Orders before the ICs had specific timelines. The timelines were and have been replaced by "immediate" and "without delay."

(5) It was stated many times throughout the rulemaking process that the NRC does not have specific threat (organization, place, or time). It was stated that there is a general threat meaning that there is intelligence to indicate that groups are interested in using radioactive material in a terrorist attack.[27]

Characteristics of the Rule

Should there be a design basis threat? NRC uses a design basis threat (DBT) to define the type of threat that a facility, such as a nuclear power plant, must be able to counter. Radiation professionals interviewed for this report had mixed views on a DBT for radioactive materials. Some felt a DBT would have been inappropriate or unworkable; one felt the rule should have had a DBT to indicate what is required; another felt that NRC has specified the threat clearly; and one noted that having a DBT for less-than-huge quantities of radioactive material would imply that there should be a DBT for chemical tanker trucks and other potential hazards.

An industrial radiographer pointed to difficulties with a DBT:

> The regulation is right not to specify a design basis threat (DBT). I'm aware of one source that has been stolen in the past 20 years. Should I hire people with M-16s to guard each of my radioactive sources? What level of security is needed? We do want operational security, but can't bust the bank for it.

A university RSO also pointed to difficulties:

> Requiring a licensee to develop a design basis threat would not work. How would a licensee know and guard against the DBT for each different Category 1 or 2 device? Requiring armed guards at every irradiator or other source would put research and medical licensees out of business for use of these devices. Also, if information on a DBT leaked out, it could reveal to terrorists what sort of attack was being defended against, and would imply that a somewhat greater threat would succeed.

An RSO at another university felt that the rule should have had a DBT:

> there is no design basis threat or discussion of adversary capabilities provided to the licensee. Without a common security education, each site will develop their security in accordance of their understanding of the regulations AND their guess of the potential threat. This variation in security implementation combined with the variation in inspection practices will produce widely varying security levels and practices.

An Agreement State official felt that a DBT could not have been tailored to each facility's characteristics:

> I don't think there has to be a design basis threat (DBT) for each facility. When people look at a facility to see how to harden it, they are in effect doing a DBT, as they are looking for vulnerabilities specific to each room. The [city] Police Department

> conducts security inspections of facilities. ... In my experience, when an organization needs to design a security system, RSOs consult with security professionals. This approach guards against a plausible threat that is specific to each room or facility, rather than a DBT. This is a better approach.

NRC commented, "The LLEA does not have the authority to conduct security inspections of the licensees under either the orders or the Part 37 rule." [28]

A university RSO maintained that NRC has clearly stated the threats that must be countered even without specifying a DBT:

> The rule does not specify a design basis threat but NRC has clearly communicated what a security program needs to do, potential scenario and the timeframes in which a police response is required. ... the NRC has included security assessments into the Increased Controls Orders and the new 10 CFR 37. Even without prescribing a DBT for materials licensees, [the RSO] believes that NRC has been crystal clear in what threats must be countered ... [and] does not see the need for additional guidance or a DBT.

An Agreement State official found a DBT inappropriate for most licensees, and questioned what other potential threats might require a DBT if one were set for most radioactive sources:

> A Design Basis Threat (DBT) is easy to construct for a nuclear power plant. There could be a DBT for a panoramic irradiator (with perhaps hundreds of thousands of curies) or for manufacturers and distributors of radioactive material, but it's not plausible to apply a DBT to other radioactive sources. Licensees having few employees and modest amounts of radioactive material could not handle a DBT. Terrorists could seize such sources, or mobile sources. But they could also attack tanker trucks carrying chlorine or gasoline, many of which are on the roads; should there be a DBT for them, too?

Is the rule too prescriptive, or too performance-based? A difficult issue in formulating 10 C.F.R. 37 was striking the right balance between a performance-based vs. a prescriptive rule. The former would specify the desired result; the latter, how the result is to be achieved. A performance-based rule would specify what a program must be able to do, such as detecting attempted theft of radioactive material and summoning a prompt police response, while leaving to each licensee how to accomplish these tasks. In contrast, a prescriptive rule would specify in detail what constitutes adequate security, such as having an iris scanner for access control and a TV camera and radiation monitor with direct links to the police central alarm station. In general, radiation professionals felt that the rule strikes the right balance, and noted that a companion "implementation document," to be issued by NRC along with the final rule, will provide more details on what NRC views as constituting compliance. Closely related is the question of whether the rule is "one size fits all," i.e., does not make allowances for differing types of licensees holding vastly different quantities of radioactive material under differing security conditions. There were several criticisms of the rule on this point.

GAO, in testimony of March 2012, found the security orders to be performance-based, resulting in uneven levels of security from one facility to the next.

> NRC's security order and implementation guidance are broadly written and do not prescribe the specific steps that licensees must take to secure their sources. Rather, they provide a general framework for what constitutes adequate security practices. ...It is up to the licensee to determine, for example, if security cameras are necessary or what types of locks or alarms are needed to secure doors or windows. ...The ability to tailor security to a facility's needs and resources is particularly important for commercial facilities with limited resources. For example, officials from smaller medical facilities told us that implementing specific security requirements—such as cameras and other surveillance equipment—could jeopardize their continued operations because of the costs associated with this equipment. NRC officials told us that given factors such as diverse economic conditions, facility type, layout, and operations of facilities, a "one size fits all" approach is neither practical nor desirable.
>
> We found that the NRC controls have been implemented in a variety of ways in the hospitals and medical facilities we visited ... At some locations, the controls resulted in significant security upgrades, such as the addition of surveillance cameras, upgrades to locks on doors, and alarms. In contrast, we observed minimal security in other facilities.[29]

The radiation professionals interviewed for this report felt that the diversity in types of organizations that use radioactive materials made a prescriptive approach impossible to implement by regulation, though the implementation document will help clarify what constitutes compliance with a performance-based regulation. Some felt that the rule is one size fits all, despite what NRC told GAO, and that that was bad policy. On the other hand, those who noted the flexibility inherent in a performance-based regulation would seem to contradict those who said that the regulation was one size fits all.

An industrial radiographer:

> Regarding the issue of whether Part 37 is too prescriptive or too performance-based, I like the rule. While I don't like some of the wording, it leaves us a lot of room to work. For example, not telling me how to do security zones is fine.

An Agreement State official:

> The new rule addresses a very diverse community possessing various radioactive materials. The rule must allow flexibility for the differing business models to comply with the rule. Based on this belief the rule is not too broadly written. The Agreement States and the USNRC have more than a decade of experience in performance-based regulation. Likewise, the regulated community has the same amount of experience complying with regulations through performance-based policies, procedures and programs. The new rule "tightens" up some of the initial orders with prescriptive requirements, while allowing performance-based implementation across a very diverse regulated community.

An Agreement State official:

> Some RSOs wanted a prescriptive order, but I told them that I didn't know the specifics of each facility, so they would need to develop a security system and RCP would then inspect it. If we found weaknesses in their system, we would not issue a violation notice but rather work with them to develop a corrective action plan. In practice, many RSOs would develop a plan, then call me to ask my opinion of the plan. ... I think

the rule strikes the right balance between being prescriptive and being performance-based.

A university RSO

> feels that the balance between the rule being prescriptive vs. performance-based has worked out well. Many stakeholders have lobbied NRC for more performance-based requirements in the rule, which is what NRC provided in this case. However, there is now complaint by some that they would like the rule to be more prescriptive. NRC and many other federal agencies have done an excellent job of providing detailed guidance on how to implement a rule in addition to the rule itself. NRC's draft guidance from 2010 for the draft of 10 CFR 37 provided a clear explanation of what the rule meant. As he understands it, NRC intends to publish supplemental guidance documents based on the experiences from the increased controls program to assist with implementation of the new rule. These guidance documents provide very useful information for licensees to successfully implement the regulation.

An Agreement State official:

> Compared to the orders, the final rule is much more prescriptive. ... it is a one size fits all set of regulations. The orders were issued incrementally to licensees needing different levels of security, such as panoramic irradiators, manufacturers and distributors of radioactive sources, and other organizations such as hospitals. The rule mandates that all licensees have the same high level of security; that may not be necessary. Security should have been tailored to users, as the orders were. For example, a licensee with a half-dozen employees using soil moisture and density gauges, which have millicuries of radioactive material, do not need the same level of security as a panoramic irradiator, which may have hundreds of thousands of curies.

A university RSO:

> The rule is one size fits all: security plans, training, procedures, new concepts (like security zones). There were orders, and we implemented them. After we had been inspected a few times, we were able to understand how to comply with the orders. Now with the new rule, we will have to set the rule as the new baseline. We will be reinspected against the new standards. It will take significantly more time. There are 37 Agreement States plus NRC, so there will be 38 interpretations of what the rule means. We will have to start all over again.

In comments in January 2011 on the draft rule, the Advisory Council on the Medical Uses of Isotopes, an official advisory body to NRC, stated,

> The ACMUI understands the need to quickly develop and implement the Increased Controls License Orders required the NRC to establish a one-size-fits all model for all types and uses of Category 1 and 2 sources. The ACMUI is concerned that the proposed Part 37 builds off of and expands the requirements of a one-size-fits all model.[30]

Other

Scarcity of resources has affected radiation security programs. Several radiation professionals noted many ways in which scarcity of resources has affected their or other programs. This raises the question of whether it will be economically feasible to secure radioactive materials to the extent, and on the timeline, that the rule requires.

A university RSO:

> The lack of resources affects inspections. For example, budget cuts reduce the number of inspectors and the time they have for each inspection.

An industrial radiographer:

> Under the rule, we must verify that the receiving licensee has a valid license. On the surface, this makes sense. But I have to contact that licensee's jurisdiction to get this information. For me, that means I have to call [the state] and ask the radiation health bureau if the license is still valid and can receive the shipment. In reality, [the state] doesn't have enough money, so that bureau is understaffed and people don't answer the phone.

NRC commented on the foregoing statement:

> Under 10 CFR Part 30, all licensees are required to verify that the recipient is authorized to possess the radioactive material. In the past, this was done by the recipient faxing the first page of their license to the sender. To prevent the use of counterfeit licenses, licensees will be required to verify the recipients' license with the appropriate regulator and the preferred way to comply with the regulation is use of the License Verification System.[31]

An Agreement State official:

> While there are no concerns about the new rule, there is the concern over the ability to promulgate the rule when required. I think all states including [mine] are experiencing that it is definitely harder to promulgate rules in this tough economic environment. Regulatory rule promulgation is perceived as anti-business and bad for the economy.

A university RSO:

> Unfortunately, at many institutions I've spoken with, the money estimated to implement the new rule is woefully inadequate.

A Nuclear Energy Institute representative:

> NRC's regulatory analysis with the final rule showed an average one-time cost of \$23,375, and an average annual cost of \$21,736, for each of the 1,400 or so licensees with category 1 or 2 sources. Most of these licensees are small business industrial radiographers, for whom these costs would be significant.

An Agreement State official:

> NRC can find a state inadequate for protecting public health and safety, and can put that state on "heightened oversight." A few years ago, [my state] was put on heightened oversight, after which more resources were devoted to the radiation control program and [the state] showed sufficient improvement to be taken off heightened oversight. The [state] radiation protection budget was cut approximately 20 percent for the 2011-12 biennium. In order to make sure the radioactive materials program has enough staff, resources are being shifted to ensure adequate support for radioactive materials.

Comments on Global Threat Reduction Initiative (GTRI). According to NRC, "The rule would impose the minimum requirements that the NRC believes are necessary to adequately protect the public health and safety and the common defense and security."[32] The National Nuclear Security Administration, a semiautonomous component of the Department of Energy, operates GTRI. The Domestic Materials Protection Program, a GTRI component, goes beyond the minimum requirements by providing security upgrades on a voluntary basis to facilities with radioactive material. To do so, it sends a team to a participating facility to assess its security situation. The team makes recommendations on security upgrades—such as iris scanners for access control, infrared cameras to provide video if the lights go out, a backup power supply, and communication links to police alarm stations—and negotiates with the facility's personnel on which to install. GTRI then contracts to have the agreed systems installed, and provides funds for maintenance for three years, after which time the facility is supposed to fund maintenance. GTRI also provides training at the Y-12 National Security Complex (TN) for law enforcement personnel,[33] and conducts training exercises at facilities around the country.

NRC and NNSA view their roles as complementary, with NRC setting minimum standards for securing radioactive material and GTRI providing a higher level of security. For example, a particularly worrisome threat is that posed by insiders. 10 C.F.R. 37 deals with this threat by requiring licensees to grant unescorted access to category 1 and 2 material only to personnel the licensee deems trustworthy and reliable. So far, this system appears to have worked satisfactorily; since 9/11, there has been only one theft of a category 2 source of radioactive material, a radiography camera with 33.7 curies of iridium-192 stolen from a truck in Texas in 2011, and that theft was not done for terrorist purposes. In contrast, GTRI installs TV cameras, tamper-alarming seals, radiation detectors, and links to police alarm stations in order to indicate a theft of radioactive material in progress by anyone, including an insider, as breaking the seal, or radiation leaking out when the material was removed, would trigger an alarm.

GTRI is not a part of 10 C.F.R. 37, and is not under NRC jurisdiction. Nonetheless, several radiation professionals interviewed for this report commented on GTRI, and their comments seem worth including. Most of them found GTRI valuable; one declined to participate on grounds of personnel time, cost of maintaining GTRI-installed equipment, and complexity of equipment.

A university RSO:

> NNSA's Global Threat Reduction Initiative installed a full suite of equipment. [The university's] reactor building also houses a 30-year-old irradiator with less than 1000 curies. The construction people ripped everything in the building apart for months in

order to install many alarms. However, the reactor people love the system, it's a great system.

An industrial radiographer:

What level of security is needed? We do want operational security, but can't bust the bank for it. NNSA's Global Threat Reduction Initiative has the best way to approach this. They would like to come out to our facility. I signed up for it, and am hopeful that as their limited budget will allow, they will come here this year or next and help us with realistic workable approaches to enhancing our security both at our fixed facilities as well as our mobile fleet.

An Agreement State official:

While it is the licensee's responsibility to develop and implement a compliant operational strategy, there are other resources available to assist them in this mission such as the Global Threat Reduction Initiative (GTRI). GTRI has performed a lot of security enhancements in [my state]. It has increased the security for many irradiators. [A] University, for example, has been very complimentary of the work that GTRI completed for them. It not only provided enhanced security devices, but also developed and implemented extensive tabletop exercises. Participants in these exercises included the FBI, the [state] Radiation Protection Section, universities, state and local response organizations, and other public and private sector stakeholders at local, state and federal levels. This exercise led to the development of many improved relationships among the participants.

As part of their voluntary security enhancement program, GTRI also provides opportunities for private and public sector security force response training at their Oak Ridge facility. This training specifically enhances the coordination of a facility's radiation safety staff and their responding security force, including LLEA. With real world staged environments in which the "players" form response plans and act them out, the value of pre-arranged cross-discipline response plans is evident and strong partnerships are formed.

An Agreement State official:

Police departments in [various] cities send all of their officers to Advanced Radiological Response Techniques training courses at the Y-12 National Security Complex in Tennessee. This is not something that [the radiation control program] mandates.

An Agreement State official:

NNSA has done a lot, including two tabletop exercises in [my state]. These have been eye-openers to emergency responders. They raised elements that the responders hadn't considered. For example, if it became necessary to evacuate 300,000 people from [city] because of a dirty bomb, where would they go? How would health workers treat them, given that emergency rooms would be flooded with people? The exercise helped the [city] learn what other assets were available to them in the event that they needed additional resources. Another exercise, at [a university], had a scenario with a two-pronged terrorist attack that stole a radioactive source and raided a reactor. This exercise gave an excellent opportunity to work through the issues and provided many lessons.

An Agreement State official:

> The Global Threat Reduction Initiative has been in [my state] a half-dozen times. They will install security systems for one of the state's larger licensees this summer. The program opens the eyes of licensees to security, and provides good training not only for the licensee, but also for the state inspectors who sit in on the training.

On the other hand, a university RSO does not use GTRI:

> I have thus far declined to use GTRI because of my organizations' concerns with the level of complexity with the security systems, their lack of personnel time to devote to this GTRI implementation, and uncertainty of the funds which will be needed to support maintaining the security systems after the three years in which GTRI provides that maintenance support.

Should trustworthiness and reliability (T&R) screening be enhanced? 10 C.F.R. 37 would require every individual granted unescorted access to category 1 or 2 sources to be deemed T&R by his or her employer. Some radiation professionals interviewed for this report were comfortable with their T&R procedures. They feel that their organizations are competent to adjudicate T&R decisions, are able to use outside investigators as needed, and have experience in adjudicating T&R through implementing the various security orders. They see their employees every day and can spot changes in behavior, and are aware that erroneously granting someone T&R status could ruin their organization. Others expressed concerns: how to determine who should adjudicate T&R decisions (a human resources specialist, an RSO, or a committee); standards used in making T&R decisions that varied from one organization to the next; the possibility of being sued by an individual not hired because he or she was denied T&R status; lack of clear standards in 10 C.F.R. 37 for denying T&R status; the difficulty of making T&R decisions based on incomplete information, such as for foreign students; and inadequate training of individuals who make T&R determinations for evaluating background information.

It can be readily argued that the current system is satisfactory, given that there has been only one case of a theft of a category 2 source of radioactive material since 9/11. On the other hand, the issue is potentially serious: one individual erroneously deemed T&R could obtain material for a "dirty bomb" that could contaminate several square miles of a city and cost many billions of dollars in cleanup costs, relocation costs, and lost productivity, as well as societal disruption.

Should there be a desire to enhance T&R adjudication, several measures might be taken:

- NRC could make T&R standards clearer. For example, the State Department's "Adjudicative Guidelines for Determining Eligibility for Access to Classified Information" provide a detailed set of criteria for this purpose. The department notes, "The ability to develop specific thresholds for action under these guidelines is limited by the nature and complexity of human behavior. The ultimate determination of whether the granting or continuing of eligibility for a security clearance is clearly consistent with the interests of national security must be an overall common sense judgment based upon careful consideration of the following [thirteen] guidelines, each of which is to be evaluated in the context of the whole person."[34] The text presents these guidelines, along with a detailed discussion of security concerns and mitigating factors for each. Standards such as these could be used even if not in the context of granting a security clearance.

- The Federal Select Agents Program is operated by the Centers for Disease Control and Prevention (CDC), part of the Department of Health and Human Services, and the Animal and Plant Health Inspection Service (APHIS), part of the Department of Agriculture. Its registry program "oversees the activities of possession of biological agents and toxins that have the potential to pose a severe threat to public, animal or plant health, or to animal or plant products." [35] It may offer another model for adjudicating T&R decisions. A university RSO said,

> One alternative approach to the current Increased Controls T&R process is the one used for a bioterrorism security risk assessment to vet people for access to select agents or toxins. Under this program, an organization's biosafety officer (BSO) sends in an application for an individual to APHIS or CDC to obtain a unique identifying number (UIN). The BSO submits a completed FD-961 form[36] with the UIN and two fingerprint cards to the FBI. The FBI does a background check, and the FBI and APHIS or CDC make the decision and give the biosafety officer a yes or no decision, all for no charge.[37]

NRC commented:

> The FBI's Criminal Justice Information Service (CJIS) performs a fingerprint-based Security Risk Assessment (SRA) for access to select agents, transmits the result, which indicates whether an individual is authorized or restricted, and maintains a database of their determination. CJIS bases their determination primarily on whether the criminal history record indicates that the person has committed any of the offenses listed in section 817 of the USA PATRIOT ACT, that would make them a "restricted person". The Department of Justice is specifically authorized to perform this function by the Public Health Security and Bioterrorism Preparedness and Response Act of 2002. CJIS transmits the result of the SRA to the CDC, which in turn sends an authorization or denial letter to the entity's Responsible Official. The Responsible Official is ultimately responsible for controlling access to the agent. A legislative change would be required to apply a similar process to users of radioactive materials.[38]

- NRC could provide training for personnel making T&R decisions.
- Difficult cases could be resolved by denying an individual unescorted access to category 1 or 2 material. Such cases do arise. It may prove hard to do an adequate background investigation of foreign nationals because information may be incomplete or available in a form that is hard for U.S. personnel to evaluate. Some U.S. citizens may have committed a serious crime many years earlier, but have had a clean record since.
- A federal agency, such as NRC, the FBI, or the Office of Personnel Management, could adjudicate difficult cases. Referring such cases to a federal agency would permit a more thorough background investigation, could apply clear standards to the results, and could reduce the risk that a person denied T&R would sue the requesting organization.
- A federal agency could adjudicate all cases. NRC commented, "This would require a legislative change."[39]

- Individuals granted T&R status could be required to have a Secret clearance. The definition of Secret information is: "'Secret' shall be applied to information, the unauthorized disclosure of which reasonably could be expected to cause serious damage to the national security that the original classification authority is able to identify or describe."[40] In its regulatory analysis for the new rule, NRC states that in issuing several orders between 2003 and 2006 for securing radioactive material, "NRC noted that a deliberate malevolent act by an individual with unescorted access to these materials has a potential to result in significant adverse impacts to the public health and safety or the common defense and security and, thus, necessitated expeditious implementation of additional fingerprint requirements."[41] Yet an individual who stole certain radioactive material could take it to the roof of a building and disperse it in minutes using explosives, potentially inflicting large costs and societal disruption. While this release would not be "information," it "reasonably could be expected to cause serious damage to the national security."

NRC commented:

> The 10 CFR Part 37 rule is being issued under the authority to protect public health and safety. "Secret" clearances apply to protecting the common defense and security. This recommendation would cause a change in the underpinnings of the 10 CFR Part 37 regulations and would result in a difference of costs to the licensees and regulators and change to the regulatory burden. ... additional analysis would be needed (and possibly a legislative change) to implement such a recommendation.[42]

Measures such as these would arguably provide a higher level of assurance that individuals granted T&R status were in fact trustworthy and reliable. At issue is whether they are needed; if so, if they are worth any added cost they might incur; and whether Congress would wish to make the requisite legislative changes.

Wrap-Up

Several general points emerge from the foregoing analysis.

- While the rule imposes many requirements on licensees, the previously-issued security orders did so as well, so the incremental burden of the rule is much less than would have been the case had the orders not been issued first.
- Since many actions must be undertaken by all facilities with category 1 or 2 material regardless of facility size, the rule would appear to impose a proportionately larger burden on small licensees than on larger ones.
- Layered defense—in which no layer is perfect but each reduces the likelihood of a successful terrorist attack—is a cornerstone of post-9/11 homeland security. The rule puts into effect a layered defense using personnel reliability, ability to detect intrusion, police response, etc. Vulnerabilities are inevitable: someone deemed T&R might not be, security systems could fail, or police might not respond in time. Nonetheless, the ability of one layer to offset weaknesses in others can be expected

to improve security, especially as terrorists would not necessarily know where vulnerabilities are or how to exploit them.

- Security can be expected to increase over time as more monitoring equipment is installed, as local law enforcement becomes more familiar with the need for rapid response, and as the T&R process becomes more effective. However, as with so many other aspects of enhanced security post-9/11, this gradual accretion of security measures has made the United States safer but cannot guarantee safety.
- Further steps might increase security further, such as a more rapid rollout of GTRI, a more uniform T&R policy, or federal adjudication of some or all T&R applications. Whether the costs of these measures are worth the potential benefit is a matter for political judgment.
- Radiation professionals who use or regulate radioactive material every working day express a wide range of views on how changes from orders to rule and details of each section of the rule affect them. NRC consulted with them and made many changes in moving from the draft to the final rule in response to their recommendations. Nonetheless, judging from comments presented throughout this report, further changes—whether to the rule or the accompanying guidance document—may prove desirable. Such changes would benefit from further consultation with radiation professionals.

APPENDIX A. DEFINITION OF "BYPRODUCT MATERIAL"

10 C.F.R. 37 regulates "byproduct material." This appendix presents the definition of the term as contained in 10 C.F.R. 30, "Rules of General Applicability to Domestic Licensing of Byproduct Material," at 10 C.F.R. 30.4, "Definitions."[43]

Byproduct material means—(1) Any radioactive material (except special nuclear material) yielded in, or made radioactive by, exposure to the radiation incident to the process of producing or using special nuclear material;

(2)(i) Any discrete source of radium-226 that is produced, extracted, or converted after extraction, before, on, or after August 8, 2005, for use for a commercial, medical, or research activity; or

(ii) Any material that—

(A) Has been made radioactive by use of a particle accelerator; and

(B) Is produced, extracted, or converted after extraction, before, on, or after August 8, 2005, for use for a commercial, medical, or research activity; and

(3) Any discrete source of naturally occurring radioactive material, other than source material, that—

(i) The Commission, in consultation with the Administrator of the Environmental Protection Agency, the Secretary of Energy, the Secretary of Homeland Security, and the head of any other appropriate Federal agency, determines would pose a threat similar to the threat posed by a discrete source of radium-226 to the public health and safety or the common defense and security; and

(ii) Before, on, or after August 8, 2005, is extracted or converted after extraction for use in a commercial, medical, or research activity.

APPENDIX B. RADIONUCLIDES AND QUANTITIES OF CONCERN REGULATED BY NRC

Quantities Correspond to Category 2 Sources in IAEA Code of Conduct

Radionuclide	Quantity of Concern, in Units of ...			Threshold (Ci)
	Terabecquerels (TBq)	curies (Ci)	grams (g)	to contaminate 1 square km assuming perfect dispersion
Americium-241	0.6	16	4.73	78
Americium-241/beryllium	0.6	16	~4.73	~78
Californium-252	0.2	5.4	0.01	49
Curium-244	0.5	14	0.17	130
Cobalt-60	0.3	8.1	0.007	11
Cesium-137	1	27	0.31	42
Gadolinium-153	10	270	0.08	390
Iridium-192	0.8	22	0.002	100
Promethium-147	400	11,000	11.66	410,000*
Plutonium-238	0.6	16	0.95	220
Plutonium-239/beryllium	0.6	16	16.22*	220*
Radium-226	0.4	11	10.93	13
Selenium-75	2	54	0.004	150
Strontium-90 (yttrium-90)	10	270	1.98	200
Thulium-170	200	5400	0.033*	2000
Ytterbium-169	3	81	0.003	600

Source: The list of radionuclides and the TBq column are from "Table 1: Radionuclides of Concern," in U.S. Nuclear Regulatory Commission, "Order Imposing Increased Controls (Effective Immediately)," EA-05-090, November 14, 2005, http://www.nrc.gov/security /byproduct/table1. pdf. (1 Bq = 1 radioactive disintegration of an atom per second; 1 Ci = 3.7 x 10^{10} disintegrations per second; 1 TBq = 27.027 Ci) NRC extracted the list and TBq column from International Atomic Energy Agency, "Code of Conduct on the Safety and Security of Radioactive Sources," Table 1, "Activities Corresponding to Thresholds of Categories," p. 16. Specific activity, used here to calculate grams, is from U.S. Department of Energy. Office of Environmental Management. "Table B.1. Characteristics of Important Radionuclides," http://www.orau.org/ptp /PTP%20Library/library/DOE/Misc/ Table%20B_1_%20 Characteristics%20of%20Important%20 Radionuclides.htm. CRS calculated columns for Ci and g. Data for column, "Threshold (Ci) to contaminate 1 square km," are from Sandia National Laboratories, *Radioactive Material Downselection and Source Prioritization Methodology: A Sandia National Laboratories Study in Support of the Global Threat Reduction Initiative,* May 8, 2009, p. 40, "Threshold Quantities Comparison." This document is Official Use Only; these figures are unclassified when not associated with a specific weapon. William Rhodes of Sandia National Laboratories provided data for cells marked with an asterisk.

Notes: "Radionuclides of concern" are those for which, in the specified "quantities of concern," NRC requires enhanced security, such as access control, personnel security, and record-keeping. These quantities are Category 2 sources in the IAEA Code of Conduct. The threshold for Category 1 sources is 100 times that for Category 2 sources; the threshold for Category 3 sources is one tenth that for Category 2 sources. A quantity of concern is a very small amount. One ounce is 28.35 grams; many quantities of concern are less than 1 gram.

"Threshold to contaminate 1 square km" shows the amount of material, in curies, to contaminate that area to a level that a person in that area for a year would receive a dose of 2 rem in the first year after an attack, the EPA/FEMA protective action guide for relocation. (The column for curies is presented to two significant figures.) NRC explained the rationale for the area chosen, 1 square km: "Given all the uncertainties, it was a criterion used that might represent significant economic losses, primarily from decontamination and disposal from cleanup efforts. The thresholds being used for significant [radiological exposure devices] and RDDs are the IAEA Code of Conduct Category 2 values." (Comments prepared by NRC, November 30, 2010.) The figures in this column assume perfectly even dispersion of material over the total area. They are a useful metric for comparing the ability of different radionuclides to contaminate, but perfect dispersion would not occur in the real world. Further, the masses of material needed to produce this level of contamination would be somewhat higher than shown because materials used in commerce would not be pure.

For 14 of the 16 isotopes, the quantity of concern (in curies) in the center column is less than the amount of material (in curies) needed to contaminate 1 square km in the rightmost column. One of the two isotopes for which this is not the case, thulium-170, is very rare in commerce, and for the other, strontium-90, the difference between the two quantities is not great. Thus, protecting quantities of concern generally suffices to protect quantities that could be used to create a "significant" RDD.

The Code of Conduct uses TBq as the benchmark to define quantities of concern; CRS converted TBq to Ci. Entry for grams is obtained by dividing TBq by specific activity (expressed in TBq/gram). Entries in right two columns for strontium-90 (yttrium-90) are for strontium-90.

*Rhodes notes that the figure for promethium-147 is so large because that isotope is "essentially a weak pure beta emitter." That is, curies measure the number of disintegrations per second, not energy emitted per disintegration. Since each disintegration of promethium-147 produces very little energy, and in a form of particles that travel only a short distance, it takes a large amount to contaminate 1 square km to the level that would produce the specified dose.

This table is from CRS Report R41890, *"Dirty Bombs": Technical Background, Attack Prevention and Response, Issues for Congress*, by Jonathan Medalia.

APPENDIX C. NUCLEAR REGULATORY COMMISSION SECURITY ORDERS TO LICENSEES

In the wake of the 9/11 attacks, NRC issued orders to its licensees to enhance the security of radioactive materials. This appendix summarizes these orders; a full listing is available on the NRC website at http://www.nrc.gov/reading-rm/doc-collections/enforcement/security/index.html#6.

- Licensees of panoramic and underwater irradiators having more than 10,000 curies of material (68 FR 35458, June 13, 2003).44 [45] NRC, under its common defense and security authority, issued this order to NRC and Agreement State licensees.
- Manufacturers and distributors of radioactive material (69 FR 5375, February 4, 2004). NRC, under its common defense and security authority, issued this order to NRC and Agreement State licensees.
- Licensees transporting radioactive materials in quantities of concern (69 FR 44407, August 2, 2005), with category 1 as the threshold for a quantity of concern. NRC, under its common defense and security authority, issued this order to NRC and Agreement State licensees.
- Increased Controls (IC) order for licensees authorized to possess 16 types of radioactive material above specified "quantities of concern." (70 FR 72128, December 1, 2005). [46] NRC acted because "the Commission has determined that certain additional controls are required to be implemented by Licensees to supplement existing regulatory requirements" to enhance the security of certain

quantities of certain radioactive sources.[47] The order required licensees to: (1) allow only "trustworthy and reliable" personnel to have unescorted access to category 1 and 2 material; (2) have a program to monitor and immediately respond to unauthorized access; (3) follow certain procedures in shipping category 2 material; (4) obtain from NRC an order to implement certain added security measures to ship category 1 material; (5) protect mobile or portable radioactive sources; (6) retain documentation; and (7) protect information describing protection of radioactive material from unauthorized disclosure. NRC issued the IC order to its licensees under its public health and safety authority. The Agreement States issued legally binding requirements, identical to the IC order, to their licensees.

- Fingerprinting orders: The Energy Policy Act of 2005 (P.L. 109-58, August 8, 2005), Section 652, amended Section 149 of the Atomic Energy Act to require fingerprinting and an FBI criminal history records check of any individual permitted unescorted access to certain radioactive materials. NRC determined that individuals with unescorted access to category 1 or 2 material required fingerprinting and a records check, and issued orders requiring these measures to irradiator licensees and manufacturer and distributor licensees (71 FR 63046, October 27, 2006), licensees making shipments of category 1 material (71 FR 62302, October 24, 2006), and to all other licensees with category 1 or 2 quantities of material (72 FR 70901, December 13, 2007). NRC issued the fingerprinting orders to its licensees under its public health and safety authority. The Agreement States issued legally binding requirements, identical to the fingerprinting orders, to their licensees.
- Certain Licensees Requesting Unescorted Access to Radioactive Material; Order Imposing Trustworthiness and Reliability Requirements for Unescorted Access to Certain Radioactive Material (Effective Immediately), (75 FR 160, January 4, 2010): "Starting in December 2007, the NRC and the Agreement States began issuing additional Orders or other legally binding requirements to the IC Licensees, imposing the new fingerprinting requirements. In the December 2007 Fingerprinting Order, Paragraph IC 1.c of the IC requirements was superseded by the requirement that 'Service provider Licensee employees shall be escorted unless determined to be trustworthy and reliable by an NRC-required background investigation.' However, NRC did not require background investigations for nonM&D [manufacturer and distributor] service provider Licensees. Consequently, only service representatives of certain M&D Licensees may be granted unescorted access to the radionuclides of concern at an IC Licensee facility, even though non-M&D service provider Licensees provide similar services and have the same degree of knowledge of the devices they service as M&D Licensees. To maintain appropriate access control to the radionuclides of concern, and to allow M&D Licensees and non-M&D service provider Licensees to have the same level of access at customers' facilities, NRC is imposing trustworthiness and reliability requirements for unescorted access to radionuclides of concern, as set forth in this Order."[48] "The Order was issued to NRC and Agreement State licensees by the NRC under its common defense and security authority."[49]

End Notes

[1] A recent example is U.S. House. Committee on Homeland Security. Subcommittee on Counterterrorism and Intelligence. "WMD Terrorism: Assessing the Continued Homeland Threat." 112th Congress, 2nd Session, November 15, 2012, http://homeland.house.gov/hearing/subcommittee-hearing-wmd-terrorism-assessing-continued-homelandthreat

[2] Congressional staff who would like the edited transcripts of these interview notes may obtain them from the author.

[3] Nuclear Energy Institute, "About NEI," http://www.nei.org/aboutnei/.

[4] International Atomic Energy Agency, *Code of Conduct on the Safety and Security of Radioactive Sources,* January 2004, http://www-pub.iaea.org/MTCD/publications/PDF/code-2004_web.pdf. As of July 2012, 113 nations, including the United States, had made political commitments to follow the Code of Conduct. International Atomic Energy Agency, "List of States that have a made a political commitment with regard to the Code of Conduct on the Safety and Security of Radioactive Sources and the Supplementary Guidance on the Import and Export of Radioactive Sources," updated July 10, 2012, http://www.iaea.org/Publications /Documents/Treaties /codeconduct_status.pdf.

[5] International Atomic Energy Agency, *Code of Conduct ...,* p. 2.

[6] NNSA commissioned Sandia National Laboratories to perform a screen similar to that done by the IAEA, which produced essentially the same list of radionuclides as the Code of Conduct. Further, "The Category 1 and 2 quantities of [the 16] radioactive sources listed in the Code of Conduct are considered the most risk significant and have been the focus of [U.S.] Federal and State efforts to place tighter controls for security." Radiation Source Protection and Security Task Force, *The 2010 Radiation Source Protection and Security Task Force Report,* p. 7, http://www.nrc.gov/ security/byproduct/2010-task-force-report.pdf.

[7] National Research Council. Division on Earth and Life Studies. Nuclear and Radiation Studies Board. Committee on Radiation Source Use and Replacement, Radiation Source Use and Replacement, Abbreviated Version, Washington, DC, 2008, p. 78, http://www.nap.edu/catalog.php?record_id=11976.

[8] The Atomic Energy Act of 1947 created the Atomic Energy Commission, which was responsible for military and civilian uses of atomic energy. That act has been replaced by the Atomic Energy Act of 1954, as amended. The Energy Reorganization Act of 1974 "assign[ed] to one agency, now the Department of Energy, the responsibility for the development and production of nuclear weapons, promotion of nuclear power, and other energy-related work, and assign[ed] to the NRC [Nuclear Regulatory Commission] the regulatory work, which does not include regulation of defense nuclear facilities. "U.S. Nuclear Regulatory Commission. "Governing Legislation," section on Energy Reorganization Act of 1974, http://www.nrc.gov/about-nrc/governing-laws.html#era-1974. This website provides summaries and full text of legislation relating to NRC.

[9] For legislation governing NRC, see U.S. Nuclear Regulatory Commission, "Governing Legislation," http://www.nrc.gov/about-nrc/governing-laws.html#era-1974.

[10] For a map of Agreement States and contact information for all states, see U.S. Nuclear Regulatory Commission, "NRC:FSME – State Regulations and Legislation," http://nrc-stp.ornl.gov/rulemaking.html.

[11] U.S. Nuclear Regulatory Commission. "Physical Protection of Byproduct Material; Proposed Rule," 75 *Federal Register* 33902-33947, June 15, 2010.

[12] Ibid., p. 33904.

[13] Personal communication, Nuclear Regulatory Commission staff, November 20, 2012.

[14] U.S. Nuclear Regulatory Commission. "Regulatory Analysis for Final Rule: Physical Protection of Byproduct Material ...," December 2011, p. 1, http://www.nrc.gov/reading-rm/doc-collections/commission/secys/2011 /2011- 0170scy.pdf, pdf page 480.

[15] U.S. Nuclear Regulatory Commission, 10 C.F.R. Parts 20, 30, 32, 33, 34, 35, 36, 37, 39, 51, 71, and 73, RIN 3150- AI12, [NRC-2008-0120], December 8, 2011, Enclosure 2, pp. 13-81 (pdf pages 404-472), http://www.nrc.gov/readingrm/doc-collections/commission/secys/2011/2011-0170scy.pdf.

[16] U.S. Nuclear Regulatory Commission. *Implementation Guidance for 10 CFR Part 37, Physical Protection of Byproduct Material, Category 1 and Category 2 Quantities of Radioactive Material,* June 2010, draft for comments, http://www.aapm.org/government_affairs/documents/Part37DraftGuidanceJune2010.pdf.

[17] U.S. Nuclear Regulatory Commission. "Physical Protection of Byproduct Material," (10 C.F.R. Parts 20, 30, 32, 33, 34, 35, 36, 37, 39, 51, 71, and 73), December 2011, http://www.nrc.gov/reading-rm/doc-collections/commission/secys/ 2011/2011-0170scy.pdf. The rule is on pages 318-370 of this document.

[18] A radiation safety officer is the individual at an organization, such as a university or a hospital complex, responsible for the safety and security of materials that produce radiation, and often also responsible for the safety and security of radiation-producing equipment.

[19] Information provided by Nuclear Regulatory Commission staff, email, June 28, 2012. Emphasis added.

[20] Information provided by Nuclear Regulatory Commission, August 7, 2012.

[21] NRC prefers to use "security visits" because "security inspections" could be "interpreted to mean that the LLEA could inspect against the orders and Part 37 requirements. Only the NRC or Agreement States can inspect against the security requirements contained in the orders and Part 37 and issue a notice of violation. Licensees can coordinate or consult with the LLEAs and in fact are required to attempt to coordinate with the LLEA. LLEAs may be able to inspect under their own authority." Information provided by Nuclear Regulatory Commission, August 14, 2012.

[22] Information provided by Nuclear Regulatory Commission, August 7, 2012.

[23] Information provided by Nuclear Regulatory Commission, November 2 and 5, 2012.

[24] Information provided by Nuclear Regulatory Commission, August 15, 2012.

[25] NRC issued an order (EA-05-006) imposing requirements on transportation of radioactive material quantities of concern (RAMQC) on July 19, 2005; the order itself was not released to the public because it contained security-sensitive information. See U.S. Nuclear Regulatory Commission. Letter from Jack R. Strosnider, Director, Office of Nuclear Material Safety and Safeguards, to Holders of Material Licenses Authorized to Possess and Transfer Items Containing Radioactive Material Quantities of Concern as listed in Attachment A to Enclosure 1, Subject: Issuance of Order for Additional Security Measures on the Transportation of Radioactive Material Quantities of Concern, December 16, 2005, http://pbadupws.nrc.gov/docs /ML0533/ML053350222.pdf.

[26] Nuclear Regulatory Commission, "10 CFR Parts 20, 30, 32, 33, 34, 35, 36, 37, 39, 51, 71, and 73, RIN 3150-AI12, [NRC-2008-0120], Physical Protection of Byproduct Material," *Federal Register* notice, p. 52; pdf page 63 of http://www.nrc.gov/reading-rm/doc-collections/commission/secys/2011/2011-0170scy.pdf. [As of the date the referenced document was issued, the rule had not been published in the *Federal Register.*] The referenced page provides a detailed description of characteristics of a safe haven.

[27] Information provided by Nuclear Regulatory Commission, August 7 and November 2, 2012. Numbering is added for clarification.

[28] Information provided by Nuclear Regulatory Commission, August 7, 2012. See also footnote 21.

[29] U.S. Government Accountability Office, *Nuclear Nonproliferation: Further Actions Needed by U.S. Agencies to Secure Vulnerable Nuclear and Radiological Materials*, GAO-12-512T, March 14, 2012, pp. 16-18, http://www.gao.gov/assets/590/589345.pdf.

[30] D. Gilley, S. Langhorst, and D. Fisher, Advisory Committee on the Medical Use of Isotopes (ACMUI), "Comments on Proposed Part 37," January 5, 2011

[31] Information provided by Nuclear Regulatory Commission, August 7 and 14, 2012.

[32] U.S. Nuclear Regulatory Commission. "Regulatory Analysis for Final Rule: Physical Protection of Byproduct Material (10 CFR Parts 20, 30, 32, 33, 34, 35, 36, 37, 39, 51, 71, and 73)," December 2011, Appendix C, Regulatory Analysis, p. C-6, http://www.nrc.gov/reading-rm/doc-collections/commission/secys/2011/2011-0170scy.pdf, pdf page 531.

[33] See Y-12 National Security Complex, Nuclear and Radiological Field Training Center, "Advanced Radiological Response Techniques (ARRT) Courses," http://www.y12.doe.gov/missions/complementarywork /nrftc /arrt_courses.php .

[34] U.S. Department of State. "Adjudicative Guidelines for Determining Eligibility for Access to Classified Information," February 3, 2006, http://www.state.gov/m/ds/clearances/60321.htm.

[35] U.S. Animal and Plant Health Inspection Service and Centers for Disease Control and Prevention. "National Select Agent Registry," http://www.selectagents.gov/.

[36] This form is the Federal Bureau of Investigation Bioterrorism Preparedness Act: Entity/Individual Information; it is available at http://www.fbi.gov/about-us/cjis/bioterrorism-security-risk-assessment-form/bioterrorism-security-riskassessment-form-fd-961 .

[37] For further information on the Select Agent Program, see U.S. Centers for Disease Control and Prevention (CDC). Office of Public Health Preparedness and Response. "CDC Select Agent Program," July 2011, http://www.cdc.gov/ phpr/documents/DSAT_brochure_July2011.pdf.

[38] Information provided by Nuclear Regulatory Commission, August 7, 2012.

[39] Ibid.

[40] U.S. White House. Executive Order 13526, "Classified National Security Information," December 29, 2009, http://www.whitehouse.gov/the-press-office/executive-order-classified-national-security-information.

[41] U.S. Nuclear Regulatory Commission. "Regulatory Analysis for Final Rule: Physical Protection of Byproduct Material (10 CFR Parts 20, 30, 32, 33, 34, 35, 36, 37, 39, 51, 71, and 73)," December 2011, p. 5, http://www.nrc.gov/readingrm/doc-collections/commission/secys/2011/2011-0170scy.pdf at pdf page 5.

[42] Information provided by Nuclear Regulatory Commission, August 7, 2012.

[43] 10 C.F.R. 30.4, "Definitions," http://www.nrc.gov/reading-rm/doc-collections/cfr/part030/part030-0004.html.

[44] For information on these irradiators, see U.S. Nuclear Regulatory Commission. "Fact Sheet on Commercial Irradiators," April 2009, http://www.nrc.gov/reading-rm/doc-collections/fact-sheets/commercial-irradiators.html.

[45] U.S. Nuclear Regulatory Commission, "In the Matter of All Panoramic and Underwater Irradiators Authorized to Possess Greater than 370 Terabecquerels (10,000 Curies) of Byproduct Material in the Form of Sealed Sources; Order Imposing Compensatory Measures (Effective Immediately)," 68 *Federal Register* 35458-35462, June 13, 2003. NRC redacted specific requirements except for those on handling information.

[46] U.S. Nuclear Regulatory Commission. "Order Imposing Increased Controls (Effective Immediately)," EA 05-090 in the matter of licensees authorized to possess radioactive material quantities of concern, November 14, 2005, http://www.nrc.gov/reading-rm/doc-collections/enforcement/security/2005/ml053130218.pdf. This order was subsequently issued in 70 *Federal Register* 72128-72132, December 1, 2005.

[47] Ibid., p. 72129.

[48] Nuclear Regulatory Commission, "In the Matter of: Certain Licensees Requesting Unescorted Access to Radioactive Material ...," 75 *Federal Register* 161, January 4, 2010.

[49] Information provided by Nuclear Regulatory Commission, August 7, 2012.

In: Protection of Radioactive Material
Editors: Elijah B. Evans and Wyatt C. Johnson
ISBN: 978-1-62808-563-1

Chapter 2

IMPLEMENTATION GUIDANCE FOR 10 CFR PART 37, "PHYSICAL PROTECTION OF CATEGORY 1 AND CATEGORY 2 QUANTITIES OF RADIOACTIVE MATERIAL"*

E. Bowden-Berry, M. Cervera, P. Goldberg, S. Hawkins, M. Horn, K. Jamgochian, R. MacDougall, D. Piskura, G. Purdy, R. Ragland, N. St. Amour, F. Sturz and G. Warren

ABSTRACT

The intent of this technical report is to provide guidance on, and to assist applicants and licensees in, the implementation of Title 10 of the *Code of Federal Regulations* (10 CFR) Part 37, "Physical Protection of Category 1 and Category 2 Quantities of Radioactive Material." This document describes methods that the U.S. Nuclear Regulatory Commission (NRC) finds acceptable for implementing the regulations.

ABBREVIATIONS AND ACRONYMS

ADAMS	Agencywide Documents Access and Management System
AEA or Act	Atomic Energy Act of 1954, as amended
ALARA	as low as is reasonably achievable
CFR	*Code of Federal Regulations*
Ci	curie
DHS	U.S. Department of Homeland Security
DOE	U.S. Department of Energy

* This is an edited, reformatted and augmented version of United States Nuclear Regulatory Commission, Protecting People and the Environment, Publication No. NUREG-2155, Manuscript Completed: February 2013 Date Published: February 2013.

DOT	U.S. Department of Transportation
EPAct	Energy Policy Act of 2005
FBI	Federal Bureau of Investigation
GPS	global positioning system
HR	human resources
HSSM	highway security-sensitive material
IAEA	International Atomic Energy Agency
IC	increased controls
IDS	intrusion detection system
kg	kilogram
LLEA	local law enforcement agency
LVS	license verification system
M&D	manufacturing and distribution
NLT	no later than
NRC	U.S. Nuclear Regulatory Commission
NSTS	National Source Tracking System
OMB	Office of Management and Budget
PDF	portable document format
PII	personally identifiable information
Q&As	questions and answers
RDD	radiation dispersion device
RED	radiation exposure device
RO	reviewing official
RSO	radiation safety officer
SGI	Safeguards Information
SGI-M	Safeguards Information-Modified (handling)
T&R	trustworthiness and reliability
TBq	terabecquerel
TCN	transaction control number
TSA	Transportation Security Administration
UPS	United Parcel Service

PURPOSE AND INTRODUCTION

This document provides guidance to a licensee or applicant for the implementation of Title 10 of the *Code of Federal Regulations* (10 CFR) Part 37, "Physical Protection of Category 1 and Category 2 Quantities of Radioactive Material." It is intended for use by applicants, licensees, and the U.S. Nuclear Regulatory Commission (NRC) staff, and it will also be available to Agreement States. The approaches and methods described in this document are not requirements. The NRC considers them to be acceptable approaches and methods of complying with the requirements in Part 37. Approaches and methods that differ from those set forth in this guidance document are acceptable if they satisfy the requirements set forth in 10 CFR Part 37. Licensees are free to propose alternative ways for demonstrating compliance with these requirements.

This document provides guidance in the form of questions and answers (Q&As) for each section (and each subsection, as applicable) of the regulation. Except for definitions under 10 CFR 37.5, "Definitions," the section reference and its associated text appear in the box at the top of the page that begins each set of Q&As. A brief explanation of the rule text follows just below the box. The NRC intends the Q&As to provide guidance on the implementation of the rule language in the box. Q&As are provided for those definitions that are specific to 10 CFR Part 37.

An individual who accesses an electronic version of this guidance document can navigate more easily by clicking on the hyperlinks for each section listed in the Contents pages. Hyperlinks also appear in responses to the questions to enable the reader to go directly to source material or sections of rule text referenced in the responses.

Certain States, called Agreement States, have entered into agreements with the NRC that give them the authority to license and inspect byproduct, source, and special nuclear materials, in quantities not sufficient to form a critical mass, that are used or possessed within their borders. Any licensee or applicant, other than a Federal entity, that wishes to possess or use licensed material in one of these Agreement States should contact the responsible officials in that State for guidance on implementing the security provisions.

Subpart A — General Provisions

§ 37.1, "Purpose"
§ 37.3, "Scope"
§ 37.5, "Definitions"
§ 37.7, "Communications"
§ 37.9, "Interpretations"
§ 37.11, "Specific Exemptions"
§ 37.13, "Information Collection Requirements: OMB Approval"

§ 37.1, "Purpose"

This part has been established to provide the requirements for the physical protection program for any licensee that possesses an aggregated category 1 or category 2 quantity of radioactive material listed in Appendix A to this part. These requirements provide reasonable assurance of the security of category 1 or category 2 quantities of radioactive material by protecting these materials from theft or diversion. Specific requirements for access to material, use of material, transfer of material, and transport of material are included. No provision of this part authorizes possession of licensed material.

Explanation

These regulations establish the security requirements for the possession, use, transfer, and transportation of aggregated category 1 and category 2 quantities of radioactive material.

Questions and Answers (Q&As)

Q1: What is the purpose of Part 37?

A1: The regulations in Title 10 of the *Code of Federal Regulations* (10 CFR) Part 37, "Physical Protection of Category 1 and Category 2 Quantities of Radioactive Material," impose security requirements for the possession and use of category 1 and category 2 quantities of radioactive material. These regulations establish the objectives and minimum requirements that licensees must meet to protect against theft or diversion. The intent of these requirements is to protect the public against the unauthorized use of a category 1 or category 2 quantity of radioactive material by reducing the risk of theft or diversion of the material.

Q2: What is a category 1 or category 2 quantity of radioactive material?

A2: The U.S. Nuclear Regulatory Commission (NRC) considers category 1 and category 2 quantities of radioactive material to be risk significant, and these quantities refer specifically to 16 radioactive materials (14 single radionuclides and 2 combinations). These materials are americium-241 (Am-241), Am-241/beryllium, californium-252, curium-244, cobalt-60 (Co-60), cesium-137 (Cs-137), gadolinium-153, iridium-192 (Ir-192), plutonium-238, plutonium-239/beryllium, promethium-147, radium-226, selenium-75, strontium-90 (Sr-90) (yttrium-90), thulium-170, and ytterbium-169. Irradiated fuel and mixed-oxide fuel are not included even though they may contain category 1 or category 2 quantities of radioactive material; other regulations cover these materials. Table 1 of Appendix A to 10 CFR Part 37 provides the thresholds for category 1 and category 2 quantities of radioactive material.

Q3: How do I know when I have an "aggregated" quantity of this radioactive material?

A3: See the Q&As related to the definition of "aggregated" under 10 CFR 37.5, "Definitions," and 10 CFR 37.47, "Security Zones."

Q4: What does it mean to say that "no provision of this part authorizes possession of licensed material"?

A4: This sentence simply clarifies that 10 CFR Part 37 only establishes requirements for the security of category 1 or category 2 quantities of radioactive material. These requirements are separate from the requirements for the issuance of a license to possess these quantities of radioactive material. The regulations in 10 CFR Part 30, "Rules of General Applicability to Domestic Licensing of Byproduct Material," establish generally applicable requirements for issuance of a license to possess byproduct material. In addition, required conditions for issuance of licenses for specific uses of byproduct material appear in 10 CFR Part 31, "General Domestic Licenses for Byproduct Material," through 10 CFR Part 36, "Licenses and Radiation Safety Requirements for Irradiators," and in 10 CFR Part 39, "Licenses and Radiation Safety Requirements for Well Logging." Similarly, generally applicable requirements for the issuance of licenses to possess special nuclear material appear in 10 CFR Part 70, "Domestic Licensing of Special Nuclear Material." The list of radioactive materials of concern in Appendix A to 10 CFR Part 37 contains radioisotopes that are typically licensed under both 10 CFR Part 30 and 10 CFR Part 70. The regulations in 10 CFR Part 37 do not relieve licensees from compliance with these parts. The regulations in 10 CFR Part 37 merely establish additional requirements for the secure possession of a category 1 or category 2 quantity of these materials.

Q5: I'm a mobile radiographer licensed to possess 60-curie sources of iridium. These sources cannot easily be made into a weapon, and there are many better and more easily accessible choices for making a radiation dispersion device (RDD) or a radiological exposure device (RED). Why do I have to have a security program to secure this low-risk source?

A5: The NRC has determined that the category 2 thresholds for the radioisotopes listed in Appendix A to 10 CFR Part 37, including Ir-192, are the levels that warrant additional protection measures. In addition, an interagency task force established by the Energy Policy Act of 2005 (EPAct) concluded in its 2006 report to Congress and the President, entitled, "The Radiation Source Protection and Security Task Force Report" (Agencywide Documents Access and Management System (ADAMS) Accession No. ML062190349), that the appropriate radioactive sources were being protected. The task force also concluded that the International Atomic Energy Agency (IAEA) Code of Conduct serves as an appropriate framework for considering which sources warrant additional protection. For its 2010 report to Congress and the President (ADAMS Accession No. ML102230141), the task force conducted a reevaluation of the list of radionuclides that warrant additional security and protection. The task force found "that the Category 1 and 2 quantities remain valid for sealed and unsealed sources as the list and threshold levels of radionuclides that could result in a significant RED or RDD event and therefore warrant enhanced security and protection."

§ 37.3, "Scope"

§ 37.3(a)

Subparts B and C of this part apply to any person who, under the regulations in this chapter, possesses or uses at any site, an aggregated category 1 or category 2 quantity of radioactive material.

Explanation

This section establishes which licensees are subject to Subpart B, "Background Investigations and Access Authorization Programs," and Subpart C, "Physical Protection Requirements during Use," of 10 CFR Part 37.

Q&As

Q1: Who is covered by Subpart B, "Background Investigations and Access Authorization Programs," and Subpart C, "Physical Protection Requirements during Use"?

A1: Subpart B and Subpart C requirements apply to any licensee that possesses or uses an aggregated[1] category 1 or category 2 quantity of radioactive material at any site. This requirement includes a wide range of licensees, including the following:

- pool-type irradiator licensees
- manufacturer and distributor licensees

[1] As defined in 10 CFR 37.5, "aggregated" means accessible by the breach of a single physical barrier that would allow access to radioactive material in any form, including any devices that contain the radioactive material, when the total activity equals or exceeds a Category 2 quantity of radioactive material.

- medical facilities with gamma-ray stereotactic radiosurgery devices
- self-shielded irradiator licensees (including blood irradiators)
- teletherapy unit licensees, radiographers
- well loggers, broad-scope users
- radioisotope thermoelectric generator licensees
- fuel cycle licensees
- research and test reactors
- commercial power reactors
- some fixed gauge licensees
- licensees that ship or prepare for shipment a Category 1 or Category 2 quantity of radioactive material.

For additional Q&As on the applicability of Subpart B, see those in 10 CFR 37.21, "Personnel Access Authorization Requirements for Category 1 or Category 2 Quantities of Radioactive Material." For additional Q&As on the applicability of Subpart C, see those in 10 CFR 37.41(a).

Q2: Do 10 CFR Part 37 requirements apply to unsealed radioactive material, as well as material sealed in a source or device? For example, do the requirements apply to materials possessed by a nuclear laundry or radioactive waste processor?

A2: Yes. The regulations in 10 CFR Part 37 apply to both sealed and unsealed radioactive materials in quantities equal to, or greater than, category 2. No distinction is made between unsealed and sealed radioactive material when implementing 10 CFR Part 37 requirements. For example, a nuclear laundry or radioactive waste processing licensee must implement these requirements if it possesses radioactive material in an aggregated quantity that meets or exceeds a category 2 threshold.

Q3: How do I determine whether 10 CFR Part 37 applies to me?

A3: See the Q&As related to the definition of "aggregated" and category 1 and category 2 quantities of radioactive material under 10 CFR 37.5 and 10 CFR 37.47.

§ 37.3, "Scope" (continued)

§ 37.3(b)

Subpart D applies to any person who, under the regulations of this chapter:

(1) Transports, or delivers to a carrier for transport in a single shipment, a category 1 or category 2 quantity of radioactive material; or

(2) Imports or exports a category 1 or category 2 quantity of radioactive material; the provisions only apply to the domestic portion of the transport.

Explanation

This section establishes which licensees and activities are subject to Subpart D, "Physical Protection in Transit," of 10 CFR Part 37.

Q&As

Q1: Who is covered by Subpart D, "Physical Protection in Transit"?

A1: Subpart D requirements apply to any licensee that transports, or delivers to a carrier for transport, a category 1 or category 2 quantity of radioactive material. A licensed importer or exporter of this material must comply with Subpart D and other applicable requirements in 10 CFR Part 37 only for the portion of the shipment carried out within the borders of the United States or its territories.

Q2: Are there any modes of transportation not covered by 10 CFR Part 37?

A2: Yes. The regulations in 10 CFR Part 37 do not address air or water transport. The Federal Aviation Administration regulates the transport of radioactive material within airports and by air. The U.S. Coast Guard regulates the transport of radioactive material within ports and by waterway.

The rule also does not address "transshipments" of category 1 or category 2 quantities of radioactive material through the United States. Transshipments are shipments that originate in a foreign country, pass through the United States, and then continue on to another country.

The rule does not address the transport of spent fuel except for fuel in quantities of 100 grams (0.22 pounds) or less in net weight, exclusive of cladding or other structural or packaging material, and fuel that has a total external radiation dose rate in excess of 1 gray (100 rad) per hour at a distance of 1 meter (3.3 feet) from any accessible surface without intervening shielding. The regulations in 10 CFR Part 73, "Physical Protection of Plants and Materials," cover transportation of larger quantities of spent fuel. The regulations in 10 CFR 73.37, "Requirements for Physical Protection of Irradiated Reactor Fuel in Transit," establish requirements that apply to the transport of spent fuel quantities in excess of 100 grams that exceed the dose rate of 1 gray per hour.

§ 37.5, "Definitions"

As used in this part:

Access control means a system for allowing only approved individuals to have unescorted access to the security zone and for ensuring that all other individuals are subject to escorted access.

Q&As

Q1: Does this definition of access control also cover individuals with access to safeguards information-modified handling (SGI-M)?

A1: No. This definition covers only the control of access to security zones, which are defined below as any temporary or permanent area determined and established by the licensee for the physical protection of category 1 or category 2 quantities of radioactive material. A licensee's access control program under 10 CFR Part 37 is not required to include Safeguards Information (SGI). Although similar, the requirements for access to radioactive materials in Part 37 are not the same as the requirements in 10 CFR Part 73 for access to SGI-M. (See 10

CFR 73.23, "Protection of Safeguards Information—Modified Handling: Specific Requirements.")

Q2: Does an individual who complies with 10 CFR Part 37 requirements for unescorted access to radioactive material automatically comply with 10 CFR Part 73 requirements for authorized access to SGI or SGI-M?

A2: No. As noted in response to Q1 above, the requirements for access to radioactive materials in 10 CFR Part 37 are not the same as the requirements for access to SGI or SGI-M in 10 CFR Part 73. Therefore, an individual who complies with the requirements for unescorted access to radioactive materials in 10 CFR Part 37 does not automatically comply with the requirements for access to SGI or SGI-M. However, licensees may use the information developed from background investigations for unescorted access to radioactive material to determine the trustworthiness and reliability (T&R) of an individual for access to SGI-M, and vice versa.

> *Act* means the Atomic Energy Act of 1954 (AEA) (68 Stat. 919), including any amendments thereto.

This definition is not specific to 10 CFR Part 37.

> *Aggregated* means accessible by the breach of a single physical barrier that would allow access to radioactive material in any form, including any devices that contain the radioactive material, when the total activity equals or exceeds a category 2 quantity of radioactive material.

Q&As

Q1: Does the definition of aggregated include unsealed sources and bulk material?

A1: Yes. The definition includes radioactive material "in any form." The intent is to include all material, whether it is in the form of a source (sealed or unsealed) or in a container of some sort, such as feed material that might be used to create a source.

Q2: If I have one source with radioactive material that equals or exceeds a category 2 quantity, is that source considered "aggregated" under this definition, even if I possess no other radioactive materials?

A2: Yes. Even if the licensee's source is inside a device, it must be considered aggregated under this definition because it contains a category 2 or greater quantity of material that is accessible by the breach of a single physical barrier.

Q3: What is a physical barrier?

A3: Under 10 CFR 37.47(c)(1), a physical barrier is a natural or manmade structure or formation sufficient for the isolation of a category 1 or category 2 quantity of radioactive material within a security zone.

Q4: How does the term "bulk material" in the NRC's definition of "aggregated" align with the U.S. Department of Transportation's (DOT's) terminology for bulk packaging?

A4: The two terms are unrelated. The DOT regulation that uses this terminology applies only to the packaging and transportation of bulk material. It does not apply to the radiation safety or security of the material itself. The NRC considers bulk material to be radioactive material that is not a discrete source or a sealed source as these terms are defined in 10 CFR 30.4, "Definitions." Bulk material is radioactive material that is not encased in a capsule that is designed to prevent leakage or the escape of the material.

Q5: Does the "total activity" in the definition of aggregated refer to the activity of all the radioactive material the licensee possesses or just the activity that would be accessible by the breach of a single physical barrier?

A5: "Total activity" in this definition refers only to the activity of the radioactive material that would be accessible by the breach of a single physical barrier. A licensee may possess quantities of radioactive material with a total activity several times greater than the category 2 threshold. However, a category 2 or greater quantity is considered "aggregated" only if it is located within an area isolated by a single physical barrier. The licensee may use or store material with radioactivity in category 2 quantities or greater at several locations. This material would not be considered aggregated as long as access to each location is controlled by at least one physical barrier.

Q6: When I'm storing radioactive material, does the phrase "physical barrier" used in the definition of "aggregated" apply to each storage container or to the area within which I'm storing these containers? May I count each drum as an independent physical barrier so that all the drums together need not be considered aggregated?

A6: The answer to these questions depends on the circumstances, including the quantity of radioactive material inside the storage containers. If a storage container contains a category 2 quantity of radioactive material, the concept of a physical barrier is not a factor as applied by 10 CFR Part 37. The security zone established could be an area surrounding the single drum. If a licensee has several drums that contain radioactive material, the material in all the drums would be added together to determine if, combined, they have a category 2 or greater quantity. The drums by themselves should not be considered a physical barrier because the material could easily be removed from the security zone by tipping the drums over and then rolling them to an unsecured location. Thus, without additional barriers, material stored together in individual drums would be considered aggregated. The regulations in 10 CFR Part 37 would apply if the licensee has a category 2 or greater quantity. However, if the lids to the drums are secured (e.g., padlocked or welded) and if the drums are individually secured to the floor with bolts, chains, or cables to prevent them from being easily removed, the sources would not be considered aggregated, and, therefore, 10 CFR Part 37 would not apply.

Q7: When I'm storing a category 2 or greater quantity of radioactive material in the form of multiple mobile devices inside a locked storage container or storage area, can each of these devices be considered a "physical barrier" under the definition of "aggregated"? If so, would the devices within that locked container or storage area be considered protected by at least two "independent physical controls" and meet the additional security requirement for mobile devices under 10 CFR 37.53(a)?

A7: The answer to both questions is no. The definition of "aggregated" specifically covers a category 2 or greater quantity of radioactive material "in any form, *including any devices that contain the radioactive material* [emphasis added]." The devices themselves cannot be considered independent physical controls because, for the purpose of unauthorized removal, access to a mobile device is effectively the same as access to the material inside the device. In this case, the "physical barrier" in the definition of "aggregated" is the locked container or the locked room or cage surrounding the area within which the licensee stores these mobile devices. If this locked container, room, or cage is the only physical barrier to gaining access to the category 2 or greater quantity of material in the devices, this material is "aggregated" under this definition and, therefore, under 10 CFR 37.53(a), the devices must be protected by at least one additional independent physical control when they are not under the direct control and constant surveillance of the licensee.

Q8: When I'm transporting several devices that individually are less than a category 2 quantity, but together are more than that, does the phrase "physical barrier" used in the definition of "aggregated" apply to each device or only to the shipping container I use to transport the devices? Can I count the individual devices containing the material as barriers along with the shipping container, so that I would have more than one physical barrier and the shipment would not have to be considered aggregated?

A8: No. The concept of physical barriers does not apply to transportation. If someone steals a truck transporting sources, he or she would have access to all of the material, regardless of any shipping container barriers. Subpart D requirements apply if the licensee transports a category 2 or greater quantity of material in a single conveyance.

Q9: If I conduct my licensed operations on a sufficiently large site, can I use the physical distance between the locations of my devices or materials as a physical barrier so that I would not have to consider them aggregated?

A9: Yes. However, the necessary physical distance separating the locations in which materials or devices or both are used or stored will vary for each licensee, and the burden will be on the licensee or license applicant to show that this distance is sufficient to qualify as a physical barrier. To demonstrate compliance, the licensee would probably need to consider such site-specific factors as follows:

1. the continuous monitoring or integration of devices into the facility's operation in such a way that the removal of such a device would provoke a prompt investigation and response
2. the ease of moving its materials or devices or both
3. the availability of tools and equipment, including vehicles, to accomplish this movement
4. the likely minimum travel time from location to location
5. the delay associated with these factors compared to the delay afforded by alternative physical barriers, such as a lock and chain or a lockable cage or room.

To avoid unanticipated delays or changes in a facility's plans or operations, the licensee should consult its regulatory agency in advance about its intent to use physical distance as a physical boundary instead of waiting for an inspection to identify the issue. The licensee should also keep in mind that physical distance would not count as a physical barrier during the removal of sources from their devices for maintenance or replacement and when they are aggregated within a common physical barrier for temporary storage.

Q10: How far apart must my devices or materials be located for them not to be considered "aggregated"?

A10: No single minimum distance between locations of materials or devices can or should be prescribed in a regulation applied to all licensees' situations. As noted in the preceding Q&A, the licensee will have to identify the minimum distance that it believes should be counted as a physical barrier based on an analysis of site-specific conditions. The analysis will be subject to NRC inspection.

> *Agreement State* means any State with which the Atomic Energy Commission or the U.S. Nuclear Regulatory Commission has entered into an effective agreement under Subsection 274b of the Act. *Non-agreement State* means any other State.

This definition is not specific to 10 CFR Part 37.

> *Approved individual* means an individual whom the licensee has determined to be trustworthy and reliable for unescorted access in accordance with Subpart B of this part and who has completed the training required by 10 CFR 37.43(c).

Q&As

Q1: Is an individual approved for access to radioactive materials automatically also approved for access to SGI or SGI-M as well?

A1: No. To be approved for access to SGI or SGI-M, such an individual would have to meet the applicable requirements in Part 73 and would have to have a need-to-know SGI to perform his or her duties. Similarly, an individual approved for access to SGI or SGI-M would also need to have access to category 2 or greater quantities of radioactive material to perform job-related duties to be approved for unescorted access to these materials.

Q2: Who approves an "approved individual"?

A2: The rule requires a reviewing official (RO) designated by the affected licensee to approve an individual for unescorted access to category 1 or category 2 quantities of radioactive material. The RO must determine that the individual seeking such access is trustworthy and reliable before granting that individual unescorted access. See the definition of an RO below.

Q3: What does it mean to be "trustworthy and reliable"?

A3: See the definition of "trustworthiness and reliability" below and the Q&As pertaining to 10 CFR 37.25, "Background Investigations."

Q4: What does it mean to "complete" the training required by 10 CFR 37.43(c)? Does an individual have to pass a test to have successfully "completed" the training?

A4: No. Testing is not required; however, evidence that the trainee took and passed a reasonable test of the knowledge, skill, or ability objectives of the training is one possible measure of the likelihood that he or she will be able to carry out his or her assigned security responsibilities. No individual subject to the training requirements may be permitted unescorted access to a category 1 or category 2 quantity of radioactive material before completing the training requirements appropriate for that individual's potential job responsibilities. See the Q&As for 10 CFR 37.43(c).

> *Background investigation* means an investigation conducted by a licensee or applicant to support the determination of trustworthiness and reliability.

See the Q&As for 10 CFR 37.25, "Background Investigations."

> *Becquerel (Bq)* means one disintegration per second.

This definition is not specific to 10 CFR Part 37.

> Byproduct material means—
>
> (1) Any radioactive material (except special nuclear material) yielded in, or made radioactive by, exposure to the radiation incident to the process of producing or using special nuclear material;
>
> (2) The tailings or wastes produced by the extraction or concentration of uranium or thorium from ore processed primarily for its source material content, including discrete surface wastes resulting from uranium solution extraction processes. Underground ore bodies depleted by these solution extraction operations do not constitute "byproduct material" within this definition;
>
> (3) (i) Any discrete source of radium-226 that is produced, extracted, or converted after extraction, before, on, or after August 8, 2005, for use for a commercial, medical, or research activity; or
>
> (ii) Any material that—
>
> (A) Has been made radioactive by use of a particle accelerator; and
>
> (B) Is produced, extracted, or converted after extraction, before, on, or after August 8, 2005, for use for a commercial, medical, or research activity; and
>
> (4) Any discrete source of naturally occurring radioactive material, other than source material, that—
>
> (i) The Commission, in consultation with the Administrator of the Environmental Protection Agency, the Secretary of Energy, the Secretary of Homeland Security, and the head of any other appropriate Federal agency, determines would pose a threat similar to the threat posed by a discrete source of radium-226 to the public health and safety or the common defense and security; and
>
> (ii) Before, on, or after August 8, 2005, is extracted or converted after extraction for use in a commercial, medical, or research activity.

This definition is not specific to 10 CFR Part 37.

> *Carrier* means a person engaged in the transportation of passengers or property by land or water as a common, contract, or private carrier or by civil aircraft.

Q&As

Q1: Can a licensee be a carrier?

A1: Yes. Any person as defined in 10 CFR 30.4 may engage in transportation of radioactive materials as a common, contract, or private carrier.

Q2: What's the difference between a "carrier" and a "transporter," a term also used in this rule?

A2: No difference between the terms exists. For the purposes of this part, the NRC uses the terms interchangeably.

> *Category 1 quantity of radioactive material* means a quantity of radioactive material meeting or exceeding the category 1 threshold in Table 1 of Appendix A to this part. This is determined by calculating the ratio of the total activity of each radionuclide to the category 1 threshold for that radionuclide and adding the ratios together. If the sum is equal to or exceeds 1, the quantity would be considered a category 1 quantity. Category 1 quantities of radioactive material do not include the radioactive material contained in any fuel assembly, subassembly, fuel rod, or fuel pellet.
>
> *Category 2 quantity of radioactive material* means a quantity of radioactive material meeting or exceeding the category 2 threshold but less than the category 1 threshold in Table 1 of Appendix A to this part. This is determined by calculating the ratio of the total activity of each radionuclide to the category 2 threshold for that radionuclide and adding the ratios together. If the sum is equal to or exceeds 1, the quantity would be considered a category 2 quantity. category 2 quantities of radioactive material do not include the radioactive material contained in any fuel assembly, subassembly, fuel rod, or fuel pellet.

Q&A

Q1: How do I determine whether I possess a category 1 or category 2 quantity of radioactive material?

A1: The licensee should use the sum-of-fractions method, which is also known as the unity rule, to determine if it possesses a category 1 or category 2 quantity of radioactive material. The licensee may need to implement 10 CFR Part 37 requirements even if it does not possess any single source or single radionuclide in excess of a category 2 threshold. For combinations of materials (including sealed sources, unsealed sources, and bulk material) and radionuclides, the licensee must include multiple sources (including bulk material) of the same radionuclide and multiple sources (including bulk material) of different radionuclides to determine if the requirements apply. For the purposes of this calculation, the licensee must consider all the radioactive material that it possesses at the site in question. The licensee may use the following formula for the unity rule to determine if it must implement the 10 CFR Part 37 requirements:

$$R1/AR1 + R2/AR2 + R_n/AR_n > 1.0$$

where:

$R1$ = total amount of radionuclide 1

$AR1$ = category 2 threshold of radionuclide 1

$R2$ = total amount of radionuclide 2

$AR2$ = category 2 threshold of radionuclide 2 R_n = total amount of radionuclide n

$AR2$ = category 2 threshold of radionuclide n and so on greater than or equal to1.0

If the sum of these fractions is greater than or equal to 1, the licensee possesses at least a category 2 quantity of radioactive material, and the 10 CFR Part 37 requirements would apply if this quantity is aggregated in a single location.

The licensee would use the same unity rule formula above to determine if it has a category 1 quantity for purposes of implementing 10 CFR Part 37 requirements for the transport of such quantities. However, it would substitute the category 1 threshold value as the divisor for each radionuclide to be shipped. Category 1 thresholds in Appendix A to this report are 100 times the thresholds for category 2 quantities.

The terabecquerel thresholds are the regulatory standard and must be used in all unity rule calculations. The licensee must therefore convert any curie values in its license for sources and material to terabecquerel as follows:

n (terabecquerel) = *N* (Ci) x 0.037 terabecquerel (TBq) per curie See the examples included below.

Example 1: The licensee possesses the following materials:

- 5 TBq of Co-60 (bulk material)
- 5 TBq of Co-60 (sealed source)
- 20 TBq of Ir-192 (sealed source)

The category 2 threshold is 0.3 TBq for Co-60 and 0.8 TBq for Ir-192. Using the sum-of-fractions formula ($R1/AR1 + R2/AR2 + R_n/AR_n > 1.0$) and dividing the total quantities of Co-60 and Ir-192 by their category 2 threshold quantities produces 58 as the sum of fractions as follows:

(5 TBq Co-60 + 5 TBq Co-60)/0.3 TBq + (20 TBq Ir-192/0.80 TBq) = 58

Because this sum is greater than 1, 10 CFR Part 37 requirements apply to this licensee. At a minimum, the licensee must apply the 10 CFR Part 37 requirements to the Co-60 and Ir-192 sources because each source individually is considered a category 2 quantity. Depending on the location of the Co-60 in bulk form, the licensee may need to apply 10 CFR Part 37 requirements if it stores that bulk material with either source or if it stores all the bulk material together at one location surrounded by one physical barrier. However, if the licensee stores the bulk material in several locations, such that each location contains less than 0.3 TBq with a physical barrier between locations, 10 CFR Part 37 requirements would not apply to the bulk material.

Example 2: The licensee possesses the following materials:

- 3 TBq of Sr-90 (bulk material)
- 0.5 TBq of Cs-137 (sealed source)
- 0.1 TBq of Am-241 (bulk material)

The category 2 threshold is 10 TBq for Sr-90, 1 TBq for Cs-137, and 0.6 TBq for Am-241. Using the sum-of-fractions formula ($R_1/AR_1 + R_2/AR_2 + R_n/AR > 1.0$) and dividing the total quantities of Sr-90 and Cs-137 by their category 2 threshold quantities produces 0.97 as the sum of fractions as follows:

(3 TBq Sr-90/10 TBq) + (0.5 TBq Cs-137/1 TBq) + (0.1 TBq Am-241/0.6 TBq) = 0.97

Because this sum is less than 1, 10 CFR Part 37 does not apply to this licensee. Example 3: The licensee possesses the following materials:

- 0.5 TBq of Cs-137 (unsealed source)
- 0.3 TBq of Am-241 (sealed source)

The category 2 threshold is 1 TBq for Cs-137 and 0.3 TBq for Am-241. Using the sum-of-fractions formula ($R_1/AR_1 + R_2/AR_2 + R_n/AR_n > 1.0$) and dividing the total quantities of Cs-137 and Am-241 by their category 2 threshold quantities produces 1 as the sum of fractions as follows:

(0.5 TBq Cs-137/1 TBq) + (0.3 TBq Am-241/0.6 TBq) = 1

Because this sum equals 1, 10 CFR Part 37 requirements may apply. If the licensee stores the two sources, 10 CFR Part 37 would apply. If the licensee stores the sources apart, each one protected by at least one independent physical barrier, 10 CFR Part 37 would not apply. See also the Q&As for definition of "aggregated."

> *Commission* means the U.S. Nuclear Regulatory Commission or its duly authorized representatives.

This definition is not specific to 10 CFR Part 37.

> *Curie* means that amount of radioactive material, which disintegrates at the rate of 37 billion atoms per second.

This definition is not specific to 10 CFR Part 37.

> *Diversion* means the unauthorized movement of radioactive material subject to this part to a location different from the material's authorized destination inside or outside of the site at which the material is used or stored.

Q&A

Q1: What's the difference between diversion and theft?

A1: Both diversion and theft involve the unauthorized movement of radioactive material to an unauthorized destination. When an adversary is actively transporting an offsite shipment away from its intended destination, the two are indistinguishable. The common dictionary meaning of theft is the wrongful "taking and carrying away" or "removing" of property from its owner, require continuing possession of the stolen material. By contrast, diversion, as defined here, does not either offsite removal or continuous possession. The diverter, for example, could conceal the implicitly requiring a movement off site. Also under these definitions, theft implicitly requires diverted material or device onsite and leave it for later misuse by either the diverter or a colluding third party.

Escorted access means accompaniment while in a security zone by an approved individual who maintains continuous direct visual surveillance at all times over an individual who is not approved for unescorted access.

Q&As

Q1: Does "accompaniment" mean that the escort must always be physically present when an individual who is not approved for unescorted access is in the security zone?

A1: Generally, yes, although an escort may monitor an unapproved individual remotely (e.g., using a closed-circuit TV system) as long as the escort maintains continuous direct visual surveillance over the individual at all times. Under some circumstances, such as when a patient undergoes teletherapy, when actual physical accompaniment of such an unapproved individual is not allowed under the as low as is reasonably achievable (ALARA) principle, remote surveillance would be the only alternative. However, remote visual surveillance should not be so remote as to interfere with the licensee's ability to detect, assess, and respond immediately to an actual or attempted theft, sabotage, or diversion of radioactive materials.

Q2: Does "direct visual surveillance" mean the same thing as "line-of-sight surveillance"?

A2: No. The term "direct continuous visual surveillance" allows the use of video surveillance.

Video surveillance would be appropriate in some, but not all, cases. For example, video surveillance of patients during a treatment would be appropriate, considering the ALARA complications of having the escort physically present in the treatment room.

Q3: This definition of escorted access requires continuous direct visual surveillance only when an individual not approved for unescorted access is within a security zone. Aren't there circumstances under which this surveillance might be necessary when an unapproved individual is on site but outside the security zone?

A3: The NRC does not require the licensee to maintain direct visual surveillance outside the security zone. The licensee, however, may choose to escort unapproved individuals in this manner in any area of the facility that it deems appropriate. The agency requires direct visual surveillance only in cases in which an unapproved individual has access to a category 1 or category 2 quantity of radioactive material, and the security zone is effectively the area within which such access is controlled.

Fingerprint orders means the orders issued by the U.S. Nuclear Regulatory Commission or the legally binding requirements issued by Agreement States that require fingerprints and criminal history records checks for individuals with unescorted access to category 1 and category 2 quantities of radioactive material or safeguards information-modified handling.

Q&A

Q1: Which particular fingerprint orders referred to in this definition do I need to know about for compliance with this rule?

A1: The existing licensees were issued NRC orders that require them to take fingerprints and to obtain criminal history records checks of employees to determine their T&R for access to radioactive materials, SGI, or SGI-M. The intent of the rule's definition here is to include other sources of fingerprinting and criminal history records check requirements, such as license conditions issued by the NRC or an Agreement State agency. A license applicant will need to meet the fingerprinting requirements in 10 CFR Part 37.

> *Government agency* means any executive department, commission, independent establishment, corporation, wholly or partly owned by the United States of America, which is an instrumentality of the United States, or any board, bureau, division, service, office, officer, authority, administration, or other establishment in the executive branch of the Government.

This definition is not specific to 10 CFR Part 37.

> *License*, except where otherwise specified, means a license for byproduct material issued pursuant to the regulations in 10 CFR Parts 30 through 36 and 39 of this chapter.

This definition is not specific to 10 CFR Part 37.

> *License issuing authority* means the licensing agency that issued the license, i.e., the U.S. Nuclear Regulatory Commission or the appropriate agency of an Agreement State.

This definition is self-explanatory and requires no Q&A.

> *Local law enforcement agency (LLEA)* means a public or private organization that has been approved by a Federal, State, or local government to carry firearms and make arrests and is authorized and has the capability to provide an armed response in the jurisdiction where the licensed category 1 or category 2 quantity of radioactive material is used, stored, or transported.

Q&A

Q1: Must an LLEA be a government entity?

A1: No. The LLEA may be any security force, including a private one, if it has Federal, State, or local government authority to carry firearms and make arrests. University police could be considered an LLEA under the definition because some university police departments are a fully badged and sworn police force with the authority to make arrests and provide an armed response. A railroad's police force with this authority also could be considered an LLEA.

See also the Q&As for 10 CFR 37.45, "LLEA Coordination and Notification," especially Q3–Q5 and A3–A5 for 10 CFR 37.45(a).

> *Lost or missing licensed material* means licensed material whose location is unknown. It includes material that has been shipped but has not reached its destination and whose location cannot be readily traced in the transportation system.

This definition is not specific to 10 CFR Part 37.

> *Mobile device* means a piece of equipment containing licensed radioactive material that is either mounted on wheels or casters, or otherwise equipped for moving without a need for disassembly or dismounting, or designed to be hand carried. Mobile devices do not include stationary equipment installed in a fixed location.

Q&As

Q1: The increased controls (IC) orders applied additional security requirements to "portable," as well as "mobile" devices. What's the difference between portable and mobile devices for purposes of compliance with 10 CFR Part 37?

A1: No difference exists from the standpoint of compliance with 10 CFR 37.53, "Requirements for Mobile Devices." The same security requirements apply to both. The definition of "mobile device" for this rule encompasses all devices that can be "hand-carried" that are mounted on wheels/casters or that are otherwise movable by an individual without additional equipment.

Q2: If I have a category 1 or 2 quantity of radioactive material in a 55-gallon drum that isn't mounted on wheels or casters and can't be carried by hand, would the NRC consider this drum a "mobile device"?

A2: Generally, no, unless the drum were light enough to be turned on its side and rolled to an unauthorized location outside an established security zone. In most cases, a loaded 55-gallon drum would likely require a handtruck, forklift, or other equipment for removal from its authorized place of use or storage. As good practice, the licensee should store the handtruck, forklift, or other equipment for moving the drum in a location away from the storage area.

Q3: If I have a category 1 or 2 quantity of radioactive material in a container that isn't mounted on wheels or casters and that cannot be carried by hand, and if I secure it in a pickup truck bed or on a trailer, would the NRC consider this source or device "mobile"?

A3: Yes. The NRC would consider the source or device mobile because it would be "mounted on wheels" and "equipped for moving without a need for disassembly or dismounting" under this definition. A container or device on a pickup truck bed or on a trailer also would need to meet the requirement in 10 CFR 37.53(b). This subsection states that unless health and safety requirements for a site prohibit disabling the vehicle, the licensee must "utilize a method to disable the vehicle or trailer when not under direct control and constant surveillance by the licensee." Licensees subject to this requirement may not rely on the removal of an ignition key to meet the requirement. (See the Q&As on 10 CFR 37.53.)

> *Movement control center* means an operations center that is remote from transport activity and that maintains position information on the movement of radioactive material, receives reports of attempted attacks or thefts, provides a means for reporting these and other problems to appropriate agencies, and can request and coordinate appropriate aid.

Q&A

Q1: Must the various functions of the "movement control center" be accomplished by one entity, or may these functions be accomplished by separate departments or personnel?

A1: The movement control center definition does not require the same department or personnel to carry out all the functions. It does require an operations center or base from which all the functions are handled. To ensure the necessary degree of coordination and effectiveness, carrying out the functions of the movement control center may be more effective under the control of a single organization managed by the individual who is responsible for the timely and effective execution of these functions. The primary purpose of the movement control center is to have staff members available who can immediately respond to an emergency and who can coordinate the required response. As specified by definition, movement control center functions must, at a minimum, include maintaining current shipment position information, receiving and sending reports, and requesting and coordinating appropriate aid.

> *No-later-than arrival time* means the date and time that the shipping licensee and receiving licensee have established as the time at which an investigation will be initiated if the shipment has not arrived at the receiving facility. The no-later-than arrival time may not be more than 6 hours after the estimated arrival time for shipments of category 2 quantities of radioactive material.

Q&As

Q1: Does the no-later-than (NLT) arrival time apply to shipments outside the United States?

A1: No. An NLT arrival time must be established only for the domestic portion of a shipment for import or export (i.e., shipments with origins and destinations within the United States or its territories or possessions). These origins may include locations at which imported materials are cleared by U.S. customs authorities, and destinations may include terminal facilities for the export of the material.

Q2: The definition of NLT arrival time specifies that the NLT arrival time may not be more than 6 hours after the estimated arrival time for shipments of category 2 quantities of radioactive material. What about category 1 quantities?

A2: The requirement for an NLT arrival time does not apply to shipments of category 1 quantities of radioactive material. For these shipments, each licensee shipping the material must establish a movement control center that must monitor shipments 24 hours a day, 7 days a week. In addition, the movement control center must have the ability to communicate immediately with the appropriate law enforcement agencies in an emergency. See the related Q&As under 10 CFR 37.75, "Preplanning and Coordination of Shipment of Category 1 or Category 2 Quantities of Radioactive Material."

> *Person* means—
>
> 1. Any individual, corporation, partnership, firm, association, trust, estate, public or private institution, group, Government agency other than the Commission or the U.S. Department of Energy [DOE] (except that the [DOE] shall be considered a person within the meaning of the regulations in 10 CFR, Chapter I, to the extent that its facilities and activities are subject to the licensing and related regulatory authority of the Commission under Section 202 of the Energy Reorganization Act of 1974 (88 Stat. 1244), the Uranium Mill Tailings Radiation Control Act of 1978 (92 Stat. 3021), the Nuclear Waste Policy Act of 1982 (96 Stat. 2201), and Section 3(b)(2) of the Low-Level Radioactive Waste Policy Amendments Act of 1985 (99 Stat. 1842)), any State or any political subdivision of, or any political entity within, a State, any foreign government or nation or any political subdivision of any such government or nation, or other entity; and
> 2. Any legal successor, representative, agent, or agency of the foregoing.

This definition is not specific to 10 CFR Part 37.

> *Reviewing official* means the individual who shall make the trustworthiness and reliability determination of an individual to determine whether the individual may have, or continues to have, unescorted access to category 1 or category 2 quantities of radioactive materials that are possessed by the licensee.

See the related Q&As for 10 CFR 37.23, "Access Authorization Program Requirements."

> *Sabotage* means deliberate damage, with malevolent intent, to a category 1 or category 2 quantity of radioactive material; a device that contains a category 1 or category 2 quantity of radioactive material; or the components of the security system.

Q&As

Q1: Must I find both "deliberate damage" *and* "malevolent intent" in order to report an incident as sabotage? Can't I presume malevolent intent if it's clear that the damage was deliberate?

A1: No. The licensee does not need to find both deliberate damage and malevolent intent to report suspected sabotage. If the licensee has determined that the damage was deliberate, it should not hesitate to report it as sabotage or vandalism, even if it has not yet found convincing evidence that the damage was malevolent. If the damage appears to have been deliberate and if it is severe enough to constitute an imminent and substantial danger to an employee or any member of the public, the licensee should report it to the LLEA immediately under 10 CFR 37.57(a), regardless of whether the damage might have been malevolent. If the licensee or the employee who discovers the damage has reason to believe that it was deliberate, the licensee should also report that it may be malevolent. In addition, regardless of any apparent malevolence, if the damage is severe enough to constitute an excessive radiation exposure to an employee or any member of the public, the licensee must report it to the NRC or appropriate Agreement State agency immediately under 10 CFR 20.2202, "Notification of Incidents." Note, too, that if a quantity of material greater than category 2 has been lost or stolen, the licensee must report it to the NRC or appropriate Agreement State agency immediately under 10 CFR 20.2201, "Reports of Thefts or Loss of Licensed Material," regardless of whether the removal of the material was malevolent.

Q2: How can I determine, for the purposes of this part, that damage to material, a device, or security system component was "deliberate"? How can I determine whether the damage was "with malevolent intent"?

A2: To answer either question, the licensee should consider one or more of the following possible questions about its discovery of the damage:

- Could such severe damage have been accidental, or could it have been accomplished only by a trained individual who knew or should have known the safety or security consequences and how to avoid them? (If the licensee concludes that the damage was not accidental, it should also consider, at a minimum, the additional questions below.)
- Have other suspicious or unexplained instances of damage to material, devices, or security equipment occurred in the past few weeks or months?
- Was the current incident immediately reported by the person who claims to have first discovered it?
- Was that person authorized to be at the site of the damage when it reportedly occurred?
- Was the damage severe enough to indicate that there was intent to degrade or disable the intended function of the damaged device or equipment?
- Does the damage affect safety or security?
- Does the damage appear to have required the efforts of more than one person?
- If the licensee is the first to discover the damage, does anything appear in the vicinity or elsewhere on the site to indicate an attempt to conceal the damage?
- Was there collateral damage to physical barriers or to monitoring or detection equipment necessary to permit successful forced access to the damaged material, device(s), or security or safety equipment?
- Did this or any other security equipment have to be defeated to gain access to the security zone?
- Is forced entry anywhere else on the premises, such as a broken window or jimmied door, evident?
- If the perpetrator(s) was discovered in the act, was he or she armed? Did he or she resist or try to escape?
- Did the incident occur at night or during any other interval when the intruder could expect reduced security capabilities at the site?

- Could the perpetrator(s) have attained access to the security zone or security equipment without the help of a knowledgeable insider?

This list of questions is not exhaustive. Other circumstances may exist that prompt other kinds of questions to determine if the damage was deliberate or the intent was malevolent for the purposes of 10 CFR Part 37.

> *Safe haven* means a readily recognizable and readily accessible site at which security is present or from which, in the event of an emergency, the transport crew can notify and wait for the local law enforcement authorities.

Q&As

Q1: Does a safe haven have to be approved by a Government entity, such as the State or local police? Do I have to get approval from anyone else, such as the owner or operator of a truck stop?

A1: No. The regulation in 10 CFR 37.75(a) only requires shipping licensees to identify safe havens during their preplanning for the shipment of a category 1 quantity of radioactive material.

Q2: Does "readily recognizable" mean that a truck transport crew has to be able to see a safe haven from the road?

A2: No. Although it would be preferable for the licensee to choose an easily recognized location, it may have to locate a safe haven in a forested area or on hilly terrain or to otherwise situate it in an area in which the physical features of the surrounding area may obscure all or part of the site. To be "readily recognizable," a safe haven only needs to be in a locale familiar enough to at least one member of the transport crew to enable him or her to know when the transport vehicle needs to turn or exit to reach the site and to know when the transport vehicle has arrived. This individual also should be able to guide local law enforcement or other assistance to the site by phone, if necessary.

Q3: Does "readily accessible" mean that the safe haven has to be along the road or right off an exit?

A3: No. Under this definition, a safe haven may be several miles from an exit or turnoff if it meets one of the definition's other criteria by having security present or by providing a site from which, in an emergency, the transport crew can notify and wait for the LLEAs.

Q4: What kind of "security" has to be "present" to satisfy this definition of safe haven? Does this mean that law enforcement agency personnel have to be onsite?

A4: Neither security personnel nor law enforcement authorities need to be present at a site for it to serve as a safe haven. Obviously, there are advantages to having security present at the site of a safe haven. However, as noted in the definition, a safe haven also may be a site from which, "in the event of an emergency, the transport crew can notify and wait for the LLEAs."

Q5: What does the NRC mean when it defines safe haven as a "site"? Does a safe haven have to have a fence or other kind of recognizable property boundaries and/or be some kind of facility, even if only a parking lot, to meet the NRC's understanding of a "site"?

A5: No. If the site is readily recognizable and readily accessible, it only needs to meet one of the remaining two criteria in the definition: (1) it must be located in an area in which security is present, or (2) it must be located in an area in which the transport crew can, in the event of an emergency, notify the LLEAs and more safely wait for them to arrive.

Q6: So, does this mean that a safe haven can be any place along the side of a road as long as there's an emergency?

A6: No, the side of the road is not typically considered a safe haven. Under 10 CFR 37.75(a)(2), as part of their preplanning prior to a shipment, the licensee must identify safe havens and coordinate shipment information with the governor or the governor's designee of any State through which the shipment will pass. A transport crew may well have no choice but to pull to the side of the road in an emergency. However, this would not be a preshipment identified safe haven. Further security is not likely to be present at these emergency stops and it may not be in an area where it is possible for the transport crew to notify LLEA and safely wait for LLEA to arrive.

Q7: What conditions enroute would constitute an "emergency" requiring a safe haven under this definition? Would the NRC consider only a security incident, such as an armed attack or other malevolent interference with a shipment, to be an "emergency"? Or could an emergency be safety-related or a mechanical problem requiring a prompt repair?

A7: An emergency could be any of these conditions and also could include severe weather or the sudden illness or incapacitation of the driver or other member of the transport crew. The key characteristic of an emergency that requires a safe haven under this definition is that the condition must result in an unexpected stoppage of the shipment. See also the pertinent Q&As for 10 CFR 37.75.

Security zone means any temporary or permanent area determined and established by the licensee for the physical protection of category 1 or category 2 quantities of radioactive material.

Q&As

Q1: What is the purpose of a security zone?

A1: A security zone is an area, defined by the licensee, that both isolates and controls access to category 1 or category 2 quantities of radioactive material to prevent and detect unauthorized access. The purpose of the zone is to define the area that contains the quantities of material that must be protected. A security zone effectively defines the area in which the licensee will apply the isolation and access control measures required by 10 CFR Part 37. Licensees must store and use all category 1 and category 2 quantities of radioactive material only within a security zone.

Q2: Are the characteristics different for "permanent" and "temporary" security zones?

A2: Both types of security zones must meet the same requirements. See the Q&As for 10 CFR 37.47; 10 CFR 37.49, "Monitoring, Detection, and Assessment"; 10 CFR 37.51, "Maintenance and Testing"; and 10 CFR 37.53.

> *State* means a State of the United States, the District of Columbia, the Commonwealth of Puerto Rico, the Virgin Islands, Guam, American Samoa, and the Commonwealth of the Northern Mariana Islands.

This definition is not specific to 10 CFR Part 37.

> *Telemetric position monitoring system* means a data transfer system that captures information by instrumentation and/or measuring devices about the location and status of a transport vehicle or package between the departure and destination locations.

Q&As

Q1: How does the NRC define "departure" and "destination" locations?

A1: Although it is not defined in the rule, "departure" marks the location at which the carrier accepts the consignment of radioactive material for shipment and begins movement of the loaded transport vehicle. The location of departure is the site that is usually, but not always, controlled by the shipping licensee and from which the movement of the loaded transport vehicle begins. "Destination" denotes the location at which the receiving licensee accepts the shipment from the carrier and unloads, or authorizes the unloading of, the radioactive material specified in the accepted shipment documents from the carrier's transport vehicle.

> *Trustworthiness and reliability* are characteristics of an individual considered dependable in judgment, character, and performance, such that unescorted access to category 1 or category 2 quantities of radioactive material by that individual does not constitute an unreasonable risk to the public health and safety or security. A determination of trustworthiness and reliability for this purpose is based upon the results from a background investigation.

Q&As

Q1: How can I or anybody really know whether an individual is "dependable" in judgment, character, and performance? What does the NRC mean by "dependable"?

A1: A judgment of any individual's dependability is inherently qualitative and not readily amenable to objective measurement or evaluation. Each individual has a temperament, personal history, and other characteristics that require a licensee's fairest subjective judgment about that individual's dependability. But the scope of this judgment is limited. For the purposes of this definition, the kind of dependability required is based on if the individual can be trusted and relied upon to comply with the licensee's security requirements to sound an alarm whenever appropriate and to do no willful harm when he or she is permitted solitary access to any location with an aggregated category 2 or greater quantity of radioactive materials. In addition, under 10 CFR 37.23(e)(1), the judgment must be based on information produced from a background investigation that satisfies the requirements in 10 CFR 37.25.

In addition, note that the basis for a judgment about an individual's dependability is limited to the information available at the time. However, if at any time between a licensee's initial investigation and the 10-year reinvestigation the licensee receives information that would call into question that individual's dependability for the purposes of this part, the licensee should not hesitate to reevaluate the basis for its earlier judgment. The regulation in 10 CFR 37.23(e)(4) specifically provides that an RO "may terminate or administratively withdraw an individual's unescorted access authorization based on information obtained after the background investigation has been completed and the individual granted unescorted access authorization."

Q2: What would constitute an "unreasonable risk" to the public health and safety or security for purposes of 10 CFR Part 37?

A2: An "unreasonable risk" in this context is a risk that, by granting the subject individual unescorted access to a category 2 or greater quantity of radioactive material, would result in theft, sabotage, or diversion of the material for a malevolent act that affects the public health and safety and the environment. See also the Q&As on 10 CFR 37.21; 10 CFR 37.23, "Access Authorization Program Requirements"; 10 CFR 37.25; 10 CFR 37.27, "Requirements for Criminal History Records Checks of Individuals Granted Unescorted Access to category 1 or category 2 Quantities of Radioactive"; and 10 CFR 37.29, "Relief from Fingerprinting, Identification, and Criminal History Records Checks and Other Elements of Background Investigations for Designated Categories of Individuals Permitted Unescorted Access to Certain Radioactive Materials or Other Property."

> *Unescorted access* means solitary access to an aggregated category 1 or category 2 quantity of radioactive material or the devices that contain the material.

Q&As

Q1: Is a patient who is alone inside the boundaries of a security zone for teletherapy, but monitored remotely, considered to have unescorted access?

A1: The patient is not considered to have unescorted access if the remote monitoring permits direct and continuous surveillance and an immediate detection, assessment, and response to any unauthorized activity. Patients undergoing radiotherapy would generally be considered "accompanied" if they are under remote, but nearby, continuous visual surveillance. These patients are also typically in rooms with radiation alarms that are closely monitored visually from an adjacent room, and they are under direct visual surveillance to permit the licensee to effectively detect, assess, and respond immediately to an actual or attempted theft, sabotage, or diversion.

Q2: When must a person who is alone inside the boundaries of a security zone, but monitored remotely, be considered "solitary"?

A2: To be considered "solitary" under this definition, the individual (1) must be unaccompanied by an approved individual while inside the boundaries of a security zone and (2) must not be under continuous direct visual surveillance at all times. (See the definition of "escorted access.") Licensees must ensure that all individuals who meet these two conditions for solitary access are approved individuals.

An unapproved individual who is solely under direct visual surveillance by an approved individual at a central monitoring station that is remote from the subject security zone should generally not be considered "accompanied" under the definition of escorted access and, therefore, should be considered sufficiently solitary to require escorting or closer surveillance. Remote visual surveillance should not be so remote that it interferes with the licensee's ability to detect, assess, and respond immediately to an actual or attempted theft, sabotage, or diversion of radioactive materials. For example, depending on the configuration of a site's walls and doors, continuous surveillance from an adjoining room could be considered "accompaniment" if the approved individual who is continuously monitoring the unapproved individual could immediately warn that individual or could summon timely assistance nearby to prevent or stop an unauthorized act. However, by this same operational criterion, continuous visual surveillance from another floor or a separate nearby building should not generally be considered "accompaniment," especially if other licensee personnel are not close enough to the unapproved individual to make a timely intervention. For the same reason, remote surveillance from another offsite facility should not be considered "accompaniment" for purposes of escorted access under similar operating conditions.

United States, when used in a geographical sense, includes Puerto Rico and all territories and possessions of the United States.

Q&As

This definition is not specific to 10 CFR Part 37.

§ 37.7, "Communications"

Except where otherwise specified or covered under the regional licensing program as provided in 10 CFR 30.6(b), all communications and reports concerning the regulations in this part may be sent as follows:

§ 37.7(a)
by mail addressed to ATTN: Document Control Desk; Director, Office of Nuclear Reactor Regulation; Director, Office of New Reactors; Director, Office of Nuclear Material Safety and Safeguards; Director, Office of Federal and State Materials and Environmental Management Programs; or Director, Division of Security Policy, Office of Nuclear Security and Incident Response, as appropriate, U.S. Nuclear Regulatory Commission, Washington, DC 20555-0001.

§ 37.7(b)
by hand delivery to the NRC's offices at 11555 Rockville Pike, Rockville, Maryland 20852.

§ 37.7(c)

Where practicable, by electronic submission, for example, Electronic Information Exchange or CD-ROM. Electronic submissions must be made in a manner that enables the NRC to receive, read, authenticate, distribute, and archive the submission and process and retrieve it a single page at a time. Detailed guidance on making electronic submissions can be obtained by visiting the NRC's Web site at http://www.nrc.gov/site-help/e-submittals.html, by e-mail to MSHD.Resource@nrc.gov, or by writing the Office of Information Services, U.S. Nuclear Regulatory Commission, Washington, DC 20555-0001. The guidance discusses, among other topics, the formats the NRC can accept, the use of electronic signatures, and the treatment of nonpublic information.

This definition is self-explanatory and requires no Q&A.

§ 37.9, "Interpretations"

Except as specifically authorized by the Commission in writing, no interpretations of the meaning of the regulations in this part by any officer or employee of the Commission other than a written interpretation by the General Counsel will be recognized as binding upon the Commission.

Explanation

This definition is self-explanatory.

Q&As

Q1: Does 10 CFR 37.9 mean I don't have to comply with an NRC regulatory requirement unless and until its meaning has been interpreted in writing by the NRC General Counsel?

A1: No. The licensee must comply with all applicable regulatory requirements in 10 CFR Part 37. This provision binds the Commission to a particular interpretation of its regulations only if the interpretation is written by the NRC General Counsel or is specifically authorized by the Commission in writing. The intent of this provision is to bind the NRC when the meaning of an NRC requirement is subject to differing and disputed interpretations. Having such a binding interpretation is not a precondition for the NRC's or Agreement State's enforcement of requirements with meanings that are plain or otherwise generally accepted. If a licensee has a question about the intended meaning of any requirement, and if this guidance does not clearly address its question, the licensee should make its question known to the NRC at its earliest convenience. The licensee may do so by contacting its nearest NRC regional office, listed in 10 CFR 30.6(b)(2).

§ 37.11, "Specific Exemptions"

§ 37.11(a)

The Commission may, upon application of any interested person or upon its own initiative, grant such exemptions from the requirements of the regulations in this part as it determines are authorized by law and will not endanger life or property or the common defense and security and are otherwise in the public interest.

Explanation

A licensee may ask the NRC for an exemption from a requirement in 10 CFR Part 37, or the NRC can unilaterally decide to exempt a licensee from a requirement. The NRC can grant exemptions as long as they are "authorized by law and will not endanger life or property or the common defense and security and are otherwise in the public interest."

Q&As

Q1: If a licensee wants to request an exemption from any requirement of the regulations, what does the licensee need to submit?

A1: A licensee can request an exemption from 10 CFR Part 37 requirements. The licensee should submit an amendment request that identifies the regulatory requirement for which it is seeking an exemption. The amendment request should explain why the licensee is requesting an exemption from the regulatory requirement and how any proposed alternatives will meet that regulatory requirement.

The NRC will evaluate the request and will determine if an exemption is appropriate. The NRC will only grant exemptions that it determines are "authorized by law and will not endanger life or property or the common defense and security and are otherwise in the public interest." It is the licensee's responsibility to provide sufficient justification in support of its exemption request to enable the NRC to make its determination.

The NRC cannot grant exemptions from requirements mandated by law. For example, the NRC cannot provide an exemption from the fingerprinting and criminal history records check for unescorted access to category 1 and category 2 quantities of radioactive material or access to SGI or SGI-M because fingerprinting and criminal history records checks are specifically required by Section 149 of the AEA, as amended by Section 652 of the EPAct.

§ 37.11, "Specific Exemptions" (continued)

§ 37.11(b)

Any licensee's NRC-licensed activities are exempt from the requirements of Subparts B and C of this part to the extent that its activities are included in a security plan required by Part 73 of this chapter.

Explanation

NRC licensees may protect the radioactive material covered by 10 CFR Part 37 under a security plan developed under 10 CFR Part 73.

Q&As

Q1: If a contract radiographer licensee with material under a 10 CFR Part 37 security plan performs work with that material at a temporary jobsite located at a reactor or fuel cycle licensee that has a security plan approved under 10 CFR Part 73, which plan would apply?

A1: The materials licensee's 10 CFR Part 37 security plan would apply for the protection of the material. The materials licensee is responsible for meeting the requirements in 10 CFR Part 37 and the licensee's own security plan at a temporary jobsite, even if the temporary jobsite is at a licensed reactor or fuel cycle facility. However, in addition to the requirements in the materials licensee's security plan, the fuel cycle or reactor licensee also may impose other requirements to meet its security plan. For example, a contract radiographer licensee who brings a source onto a reactor site to conduct radiography activities may be subject to such additional reactor security requirements as personal and vehicle searches, access control to vital areas, and training.

Q2: If a 10 CFR Part 50 reactor licensee or a 10 CFR Part 70 fuel cycle facility is also licensed under its respective license to possess an individual source, such as a radiography camera that is at or above the category 2 threshold, would 10 CFR Part 37 requirements apply for the security of the source?

A2: The answer depends on the circumstances. If the licensee protects the source under its approved 10 CFR Part 73 security plan for the reactor area, it would be exempt from the security provisions of 10 CFR Part 37. If the source is in an area not covered by the 10 CFR Part 73 security plan, the 10 CFR Part 37 provisions would apply. Similarly, for a 10 CFR Part 70 licensee, if the licensee protects the source under its 10 CFR Part 73 security plan for the fuel facility, it would be exempt from the 10 CFR Part 37 requirements. To ensure that all affected radioactive materials are covered by a security plan, the reactor or fuel cycle facility licensee should document which materials will be covered under 10 CFR Part 37 and which will be covered under 10 CFR Part 73.

Q3: Are activation products contained in, or part of, the reactor structure subject to 10 CFR Part 37?

A3: No. Activation products contained in the structure (such as the stainless steel lining of a reactor vessel, stainless steel bolts, or the reactor hull) would not be subject to 10 CFR Part 37 as long as these materials remain an integral component of a reactor. However, upon decommissioning of the reactor, waste generated from decommissioning may be subject to 10 CFR Part 37 if the waste meets or exceeds a category 2 threshold. For example, shipments of decommissioned reactor components would be subject to Subpart D if their total activity met or exceeded a category 2 threshold and if they weighed less than 2,000 kg (4,409 lbs). See 10 CFR 37.11(c) below.

§ 37.11, "Specific Exemptions" (continued)

§ 37.11(c)

A licensee that possesses radioactive waste that contains category 1 or category 2 quantities of radioactive material is exempt from the requirements of Subpart B, C, and D of this part. Except that any radioactive waste that contains discrete sources, ion-exchange resins, or activated material that weighs less than 2,000 kg (4,409 lbs) is not exempt from the requirements of this part. The licensee shall implement the following requirements to secure the radioactive waste:

§ 37.11(c)(1)

Use continuous physical barriers that allow access to the radioactive waste only through established access control points;

§ 37.11(c)(2)

Use a locked door or gate with monitored alarm at the access control point;

§ 37.11(c)(3)

Assess and respond to each actual or attempted unauthorized access to determine whether an actual or attempted theft, sabotage, or diversion occurred; and

§ 37.11(c)(4)

Immediately notify the LLEA and request an armed response from the LLEA upon determination that there was an actual or attempted theft, sabotage or diversion of the radioactive waste that contains category 1 or category 2 quantities of radioactive material.

Explanation

This section exempts radioactive wastes that contain category 2 quantities or greater of radioactive material from the security requirements contained in Subparts B, C, and D of 10 CFR Part 37. Instead, the radioactive waste is subject to the security requirements contained in 10 CFR 37.11(c)(1) through 10 CFR 37.11(c)(4). However, if the waste contains discrete sources, ion-exchange resins, or activated material weighing less than 2,000 kg (4,409 lbs), the requirements contained in Subparts B, C, and D apply.

Q&As

Q1: What radioactive material does the NRC consider "waste" for purposes of 10 CFR Part 37?

A1: As defined in 10 CFR 20.1003, "Definitions," "waste" means "those low-level radioactive wastes containing source, special nuclear, or byproduct material that are acceptable for disposal in a land disposal facility." For the purposes of this definition, "low-level radioactive waste," as defined in the Low-Level Radioactive Waste Policy Amendments

Act of 1985 (NUREG-0980, "Nuclear Regulatory Legislation: 111th Congress," Volume 1, Number 9, "2nd Session," issued January 2011, page 3-3), means "radioactive material that...is not classified as high-level radioactive waste, transuranic waste, spent nuclear fuel, or byproduct material, as defined in Section 11e.(2) of the Atomic Energy Act of 1954 [i.e., uranium mill tailings]...."

Q2: What types of radioactive wastes are exempt?

A2: Radioactive waste with diffuse category 1 or category 2 quantities, as opposed to discrete sources, is exempt only from the requirements in Subparts B, C, and D of 10 CFR Part 37. These wastes are subject to the requirements in 10 CFR 37.11(c)(1) through 10 CFR 37.11(c)(4) in Subpart A, "General Provisions," of 10 CFR Part 37. Radioactive waste that contains items, such as contaminated clothing, gloves, soil, or low specific activity waste, would be exempt from the requirements of Subparts B, C, and D. Radioactive waste that contains discrete sources, ion-exchange resins, and activated material that weighs less than 2,000 kg (4,409 lbs) are subject to the requirements of Subpart B, C, and D of 10 CFR Part 37.

Q3: When are activated material and component parts considered to be waste? If the walls (concrete, steel, etc.) become activated during use, does 10 CFR Part 37 apply?

A3: Walls and component parts that become activated throughout their life are not considered to be waste until they are no longer useful for their intended purpose. The regulations in 10 CFR Part 37 do not apply to activated material in walls and components during the operating life of a reactor, hot cell, or accelerator. Once the licensee has begun decommissioning and is dismantling the facility, the rubble and removed components would be considered waste, and 10 CFR Part 37 would then apply.

Q4: Does the 2,000-kg (4,409 lbs) limit apply to the discrete sources and ion-exchange resins or just the activated material?

A4: This weight limit only applies to activated materials and does not apply to discrete sources and ion-exchange resins.

Q5: What is meant by discrete sources? Would a waste drum that contained multiple category 3 sources that add up to a category 2 quantity of radioactive material be exempt?

A5: Consistent with 10 CFR 30.4, a discrete source means a radionuclide that has been processed to purposely increase its concentration within a material for use in commercial, medical, or research activities. Waste that has been made radioactive incidentally by a process is not considered a discrete source. The drum would not be exempt because, in this example, it contains discrete sources. The licensee would need to protect the drum containing aggregated sources under 10 CFR Part 37.

§ 37.13, "Information Collection Requirements: OMB Approval"

§ 37.13(a)

The Nuclear Regulatory Commission has submitted the information collection requirements contained in this part to the Office of Management and Budget (OMB) for approval as required by the Paperwork Reduction Act (44 U.S.C. 3501 et seq.). The NRC may not conduct or sponsor, and a person is not required to respond to, a collection of information unless it displays a currently valid OMB control number.

OMB has approved the information collection requirements contained in this part under control number 3150-0214.

§ 37.13(b)

The approved information collection requirements contained in this part appear in §§ 37.21, 37.23, 37.25, 37.27, 37.29, 37.31, 37.33, 37.41, 37.43, 37.45, 37.49, 37.55, 37.57, 37.71, 37.75, 37.77, 37.79, and 37.81.

Explanation

This definition is self-explanatory and requires no Q&A.

Subpart B — Background Investigations and Access Authorization Programs

§ 37.21, "Personnel Access Authorization Requirements for Category 1 or Category 2 Quantities of Radioactive Material"

§ 37.23, "Access Authorization Program Requirements"

§ 37.25, "Background Investigations"

§ 37.27, "Requirements for Criminal History Records Checks of Individuals Granted Unescorted Access to Category 1 or Category 2 Quantities of Radioactive Material"

§ 37.29, "Relief from Fingerprinting, Identification, and Criminal History Records Checks and Other Elements of Background Investigations for Designated Categories of Individuals Permitted Unescorted Access to Certain Radioactive Materials or Other Property"

§ 37.31, "Protection of Information"

§ 37.33, "Access Authorization Program Review"

10 CFR 37.21, "Personnel Access Authorization Requirements for Category 1 or Category 2 Quantities of Radioactive Material"

§ 37.21(a), "General"

§ 37.21(a)(1)

Each licensee that possesses an aggregated quantity of radioactive material at or above the category 2 threshold shall establish, implement, and maintain its access authorization program in accordance with the requirements of this subpart.

§ 37.21(a)(2)

An applicant for a new license and each licensee that would become newly subject to the requirements of this subpart upon application for modification of its license shall implement the requirements of this subpart, as appropriate, before taking possession of an aggregated category 1 or category 2 quantity of radioactive material.

> **§ 37.21(a)(3)**
> Any licensee that has not previously implemented the Security Orders or been subject to the provisions of Subpart B shall implement the provisions of Subpart B before aggregating radioactive material to a quantity that equals or exceeds the category 2 threshold.

Explanation

These provisions require licensees to have an access authorization program. New license applicants and licensees newly subject to 10 CFR Part 37 must implement this program before taking possession of aggregated category 1 or category 2 quantities of radioactive material. A licensee that has not previously implemented NRC security orders or has not been subject to the provisions of Subpart B must implement these provisions before aggregating radioactive material to a category 2 or greater quantity.

Q&As

Q1: Who is required to have an access authorization program?

A1: Any licensee that possesses an aggregated category 1 or category 2 quantity of radioactive material at a facility must have an access authorization program under 10 CFR 37.21(a). For first-time license applicants or those who are seeking a license amendment to authorize possession of an aggregated category 2 or greater quantity of radioactive material, this means that the access authorization program must be in place before the licensee may possess the aggregated material. It also means that if a licensee must implement a security plan under Subpart C or implement the provisions of Subpart D of 10 CFR Part 37 for such aggregated quantities, the licensee will need to have an access authorization program as well.

Q2: I do not currently possess a category 1 or category 2 quantity of radioactive material, but my license authorizes me to possess this quantity. Do I need to implement an access authorization program?

A2: No. If the licensee does not actually possess a category 1 or category 2 quantity of radioactive material, it does not have to implement an access authorization program. However, it does need to implement the program before it can take possession of that quantity and before it can aggregate the material at or above a category 2 threshold.

Q3: I am already authorized to possess, at several separate locations, radioactive material that adds up to more than a category 2 quantity, and I now need to aggregate this material at a single location. Will I need to implement an access authorization program?

A3: Yes. If a licensee plans to aggregate the material to a category 2 or greater quantity, it will need to develop and implement an access authorization program before it can aggregate the material. The licensee will need to conduct a background investigation, including fingerprinting, for anyone who will have unescorted access to the aggregated category 2 or greater quantity of radioactive material. These individuals will need to be approved for access before the material is delivered for aggregated storage at the single location. The licensee also should notify the appropriate NRC regional office that it will be implementing an access authorization program for the first time.

Q4: May I obtain an exemption from the access authorization program requirements if I implement alternative physical or administrative controls and training?

A4: A licensee may apply for an exemption from an access authorization program requirement. The Commission's discretion to grant an exemption from the access authorization requirements in Subpart B, however, is constrained by statute and a formal determination by the Commission. Under Section 149 of the AEA, the NRC must fingerprint "any individual" granted unescorted access to radioactive material subject to NRC regulation that the Commission determines to be "of such significance to the public health and safety or the common defense and security as to warrant fingerprinting and background checks." The Commission has determined that unescorted access to material in category 2 or greater quantities requires fingerprinting and background investigations. The Commission will not grant an exemption for this statutory requirement.

The requirements in 10 CFR Part 37 are designed to provide defense in depth for the security of aggregated quantities of radioactive material. No single measure can provide the same level of protection as that of all security measures integrated into a compliant program. Therefore, if the licensee will be aggregating radioactive material into a category 2 quantity or greater, it must implement each applicable 10 CFR Part 37 requirement unless it can justify the need for an exemption and can obtain NRC approval.

Q5: Will I be subject to Subpart B and need to have an access authorization program if I apply to amend my license to increase its possession limits to a category 2 or greater quantity?

A5: If a licensee plans to possess a category 2 or greater quantity of radioactive material that will be aggregated at one location, it must, under 10 CFR 37.21(a)(2), implement an access authorization program before obtaining the material. However, if the licensee plans to possess the material in less than category 2 quantities at several locations, it does not need to establish or implement an access authorization program before obtaining the radioactive material.

§37.21, "Personnel Access Authorization Requirements for Category 1 or Category 2 Quantities of Radioactive Material" (continued)

§ 37.21(b), "General Performance Objective"

The licensee's access authorization program must ensure that the individuals specified in paragraph (c)(1) of this section are trustworthy and reliable.

Explanation

This provision establishes the general performance objective of the access authorization program.

Q&As

Q1: What is the objective of the access authorization program?

A1: The main objective of the access authorization program is to ensure that individuals who have unescorted access to a category 1 or category 2 quantity of radioactive material are trustworthy and reliable and do not constitute an unreasonable risk to the public health and safety or security of the material.

Q2: How can I protect against an insider threat?

A2: The regulations in 10 CFR Part 37 require licensees to limit unescorted access to category 1 or category 2 quantities of radioactive material to approved individuals. Under 10 CFR 37.25 and 10 CFR 37.27, a background investigation that includes fingerprinting and a Federal Bureau of Investigation (FBI) criminal history records check must determine if an approved individual is trustworthy and reliable. In addition, under 10 CFR 37.43(c), the licensee must provide training to its staff. This training should, among other things, enhance its employees' and contractors' awareness of the requirements in 10 CFR 37.57(b) and 10 CFR 37.81(c) to assess and report, as appropriate, "any suspicious activity related to possible theft, sabotage, or diversion" of category 1 or category 2 quantities of radioactive material. Such activity could include unusual or suspicious behavior by employees or contractor employees with routine access to areas of the site or equipment related to the control of access to a security zone.

In addition, the rule establishes similar, but separate, controls on access to materials and access to information in security plans and procedures. Thus, a licensee may grant an approved individual unescorted access to the radioactive material while limiting or restricting that individual's access to items, such as security system codes, monitoring system configurations, alarm system capabilities, and other information about the physical protection of the material. As with access to material, access to such information must be limited to individuals with a need to know who have been determined to be trustworthy and reliable.

These complementary, but independent, sets of restrictions may not prevent a determined insider; however, they will reduce the risk of an individual with malicious intent gaining or enabling others to gain access to the radioactive material.

§ 37.21, "Personnel Access Authorization Requirements for Category 1 or Category 2 Quantities of Radioactive Material" (continued)

§ 37.21(c), "Applicability"

§ 37.21(c)(1)

Licensees shall subject the following individuals to an access authorization program:

§ 37.21(c)(1)(i)

Any individual whose assigned duties require unescorted access to category 1 or category 2 quantities of radioactive material or to any device that contains the radioactive material.

§ 37.21(c)(1)(ii)
Reviewing officials.

§ 37.21(c)(2)
Licensees need not subject the categories of individuals listed in 10 CFR 37.29(a)(1) through (13) to the investigation elements of the access authorization program.

§ 37.21(c)(3)
Licensees shall approve for unescorted access to category 1 or category 2 quantities of radioactive material only those individuals with job duties that require unescorted access to category 1 or category 2 quantities of radioactive material.

§ 37.21(c)(4)
Licensees may include individuals needing access to safeguards information-modified handling under 10 CFR Part 73 in the access authorization program under Subpart B.

Explanation

These provisions establish who is subject to the access authorization program.

Q&As

Q1: Who is subject to the licensee's access authorization program?

A1: Individuals subject to a licensee's access authorization program include ROs and anyone allowed unescorted access to a category 1 or category 2 quantity of radioactive material. Unescorted access is defined as solitary access to category 1 or category 2 quantities of radioactive material or devices that contain this material. The term applies to anyone who is unaccompanied while in a security zone by an approved individual who maintains continuous direct visual surveillance at all times. (See the Q&As for definitions of "unescorted access" and "escorted access" in 10 CFR 37.5.) This would include an unaccompanied contract carrier or other contract service provider.

The licensee may accept the background investigation conducted by other licensees, such as service providers, if it receives appropriate documentation. However, the licensee still needs to make the determination that the individual requires unescorted access at its facility and that it accepts the documentation from the other licensee. The licensee will need to verify basic identification information before it can grant unescorted access to the individual, and the individual must complete the training required by 10 CFR 37.23(a)(2).

For road shipments of category 1 quantities of radioactive material under 10 CFR 37.79(a)(1) of Subpart D, the access authorization program must also include vehicle drivers and accompanying individuals. Movement control center personnel and any individual whose assigned duties provide access to shipment information on category 1 quantities of radioactive material are subject to access authorization requirements in 10 CFR 73.21 and 10 CFR 73.23

for access to SGI. These individuals may, but are not required to, be included in the licensee's 10 CFR Part 37 access authorization program.

Q2: Why do these individuals need to be subject to the access authorization program?

A2: Individuals who have unescorted access to a category 1 or category 2 quantity of radioactive material could pose a threat to the public health and safety or security because they could divert or steal risk-significant quantities of radioactive material or because they could aid others in the commission of such acts.

Q3: Am I required to subject all my employees to the access authorization program?

A3: No. If an unapproved individual has a job-related need to enter a licensee's security zone(s), the licensee has the option to escort that individual and not make a T&R determination with respect to that individual. However, each escort and any other individual with the need for unescorted access will need to undergo fingerprinting and a background investigation under 10 CFR Part 37 or under a comparable access authorization program that meets the requirements in 10 CFR Part 37.

Q4: You note above that unescorted access is defined under 10 CFR 37.5 as "solitary" access. If an unapproved individual is alone inside the boundaries of a security zone, but monitored remotely, must he or she be considered "solitary" for purposes of the access authorization requirements of Subpart B?

A4: No. To be considered "solitary" under the definition of unescorted access under 10 CFR 37.5, the person must be unaccompanied by an approved individual while inside the boundaries of a security zone and must not be under continuous direct visual surveillance at all times. Remote visual surveillance can be considered an acceptable form of accompaniment if it is continuous and direct at all times; however, it should not be so remote that it interferes with the licensee's ability to detect, assess, and respond immediately to an actual or attempted theft, sabotage, or diversion of radioactive material. For example, depending on the configuration of a site's walls and doors, continuous surveillance from an adjoining room could be considered "accompaniment" if the approved individual continuously monitoring the unapproved individual could immediately warn that individual or could summon timely assistance nearby to mitigate the potential consequences of an unauthorized act. By this same operational criterion, continuous visual surveillance from another floor or a separate nearby building should not generally be considered "accompaniment," especially if other licensee personnel are not near enough to the unapproved individual to make a timely intervention. For the same reason, remote surveillance from another facility offsite should not be considered "accompaniment" for purposes of escorted access under similar operating conditions.

Q5: May other individuals (e.g., contract physicians, physicists, laboratory staff, housekeeping, or security staff or other staff) not actually using the device or material be authorized unescorted access to a device or radioactive material in a category 2 or greater quantity?

A5: Yes. Other personnel (both licensee and nonlicensee) may have unescorted access to the security zone in which the licensee uses or stores the device or radioactive material if it has determined that such personnel need to have such access and have been approved after undergoing a background investigation that meets the requirements of Subpart B.

Q6: Do other individuals, such as contract physicians, physicists, housekeeping, etc., need to be fingerprinted even though they don't work directly with the radioactive material?

A6: Yes. These other individuals must be fingerprinted unless their access to the security zone is always escorted. However, under 10 CFR 37.21(c), all personnel (both licensee and nonlicensee) must be fingerprinted and undergo a background investigation if they have job duties that require unescorted access to a security zone in which the licensee uses or stores materials.

Q7: If the aggregated radioactive material is in a secured area within a room, is a trustworthiness and reliability (T&R) determination still required for people who need access to that room?

A7: The answer to this question depends on how the licensee defines the security zone. If the room itself is established as the security zone, the licensee must determine if an individual with a need to work in that room is trustworthy and reliable unless he or she is escorted at all times when in the room. However, if the licensee establishes the security zone as a smaller secured area within the room, it would not have to determine if the individual is trustworthy and reliable to access that part of the room outside the perimeter of the security zone.

Q8: During a source disconnect, would an individual coming in to provide source retrieval services be subject to the access authorization program? If this service individual works for another company and [if] the company has performed a background investigation on the individual, may I accept that as adequate for my company?

AS: The answer is yes to both questions, unless the licensee escorts those individuals.

During source retrieval operations, the licensee must escort individuals without a T&R determination but with a need for access. Alternatively, the licensee may accept a service provider licensee's background investigation if it implements 10 CFR Part 37 requirements. However, the service provider must provide the licensee with a written verification of its employee's T&R under 10 CFR 37.29(a)(13). The written verification from the service provider licensee must include the name of each employee who will be providing the service and a statement that the employee has been determined to be trustworthy and reliable in accordance with the requirements in 10 CFR Part 37. The licensee must verify the identity of the employee and must document its determination basis.

Q9: Would individuals transporting radioactive material be subject to the background investigation requirements?

A9: Yes. For shipments of both category 1 and category 2 quantities of radioactive material that the licensee itself is transporting, the individuals transporting the radioactive material would be subject to the access authorization program because they would have access to the material during shipment. For road shipments of category 1 quantities of radioactive material, all vehicle drivers and accompanying individuals, whether they are licensee or contractor employees, would require fingerprinting and background investigations under 10 CFR 37.21(c)(1)(i) because their assigned job duties effectively give them unescorted access to this material. Movement control center personnel and any individual whose assigned duties provide access to category 1 shipment information that requires SGI-M handling would require fingerprinting and background checks. However, for road shipments of category 2 quantities of radioactive material, 10 CFR 37.29(a)(10) and 10 CFR 37.29(a)(11) relieve commercial drivers and package handlers, respectively, from the background investigation elements of the access authorization program. These individuals are subject to DOT security requirements.

Q10: In our operation, practically everyone knows at least something about expected category 1 shipments. Is a T&R determination required for everyone within the operation?

A10: No. However, the regulation in 10 CFR 73.23(b)(2) requires the licensee to make a T&R determination on every individual who is granted access to SGI-M. Under 10 CFR 73.23(a)(2), SGI-M specifically includes shipping information on category 1 quantities of radioactive material. Types of shipment information in 10 CFR 37.75 through 10 CFR 37.81, "Reporting of Events," that could be considered SGI-M under 10 CFR 73.23(a)(2) would include, among other things, information on physical security plans and procedures, immobilization devices, and escort requirements that are more detailed than NRC regulations; scheduling and itinerary information; arrangements with, and capabilities of, local police response forces; locations of safe havens; and details of alarm and communication systems, communication procedures, and duress codes. The performance requirements for handling SGI-M appear in 10 CFR 73.21, "Protection of Safeguards Information: Performance Requirements," and 10 CFR 73.23(a) itemizes specific types of information that the licensee should protect as SGI-M.

Q11: Is the reviewing official subject to the access authorization program?

A11: Yes. The RO must undergo fingerprinting and a background investigation. See the Q&As on 10 CFR 37.23(b) for additional information on ROs.

Q12: May additional employees (e.g., new hires or existing employees changing positions within the company) who did *not* have unescorted access under the fingerprinting orders be granted unescorted access to a category 1 or category 2 quantity of radioactive material without undergoing fingerprinting and a background investigation?

A12: No. Before being granted unescorted access to material, all employees identified by the licensee as requiring unescorted access must undergo fingerprinting and the background investigation and must be determined to be trustworthy and reliable. The licensee may escort these individuals until a background investigation has been completed.

Q13: May a reviewing official who authorizes access to radioactive materials also approve individuals for access to SGI?

A13: Yes. The regulations in 10 CFR Part 73 require that an RO conduct the background check review but they do not specify who that individual must be nor list any qualifications for the position. A licensee may choose to use the same individual for SGI access under 10 CFR Part 73 and for unescorted access to materials under 10 CFR Part 37.

Q14: May service providers not associated with a manufacturing and distribution (M&D) license be provided unescorted access to a category 1 or category 2 quantity of radioactive material at a customer facility?

A14: Yes. Service provider licensees may make a T&R determination for individuals who provide service at their customer's facilities. Service provider employees who have not been determined to be trustworthy and reliable must be escorted by an employee of the customer who is authorized to have unescorted access to the radioactive material or to the device that contains it. See the Q&As on 10 CFR 37.29(13).

Q15: Our industry is subject to three different Federal background check programs: Bureau of Alcohol, Tobacco, Firearms, and Explosives; the DOT; and the NRC. All three Federal agencies have different requirements, which can be very cumbersome, confusing, and costly. Must I establish yet another background investigation program simply to comply with 10 CFR Part 37?

A15: The need for a separate background investigation program for 10 CFR Part 37 will depend on the specific requirements of the other applicable Federal agency background investigation program(s). If a licensee has a background investigation program for other activities at its site, and if that program also complies with the requirements of 10 CFR Part 37, the licensee does not need to create a separate program for the radioactive material subject to 10 CFR Part 37. However, the licensee will need to document how it uses specific elements of existing programs to implement each 10 CFR Part 37 requirement and why it expects that these elements will demonstrate compliance with each requirement.

Q16: Do individuals on an oil rig who actually manipulate the drilling tools have to be approved individuals for access control purposes under 10 CFR Part 37 if they are not employed by the licensee?

A16: No. They do not have to be approved individuals for access control purposes under 10 CFR Part 37 as long as an approved individual escorts these individuals or exercises direct control over the devices with radioactive material.

§ 37.23, "Access Authorization Program Requirements"

§ 37.23(a), "Granting Unescorted Access Authorization"

§ 37.23(a)(1)
Licensees shall implement the requirements of this subpart for granting initial or reinstated unescorted access authorization.

§ 37.23(a)(2)
Individuals who have been determined to be trustworthy and reliable shall also complete the security training required by 10 CFR 37.43(c) before being allowed unescorted access to category 1 or category 2 quantities of radioactive material.

Explanation

Each licensee must implement the requirements in Subpart B of 10 CFR Part 37 before granting an individual unescorted access to category 1 or category 2 quantities of radioactive material. An individual must also complete security training before exercising the unescorted access.

Q&As

Q1: What additional NRC requirements must be met before allowing an individual unescorted access to a category 1 or category 2 quantity of radioactive material?

A1: In addition to the background investigation and determination of T&R, the individual must complete the security training required by 10 CFR 37.43(c) before being allowed unescorted access to the material. The individual must have an adequate understanding of his or her responsibilities. In addition, other applicable health and safety requirements may apply that the individual will have to meet before he or she is allowed unescorted access or can use the material.

§ 37.23, "Access Authorization Program Requirements" (continued)

§ 37.23(b), "Reviewing Officials"

§ 37.23(b)(1)

Reviewing officials are the only individuals who may make trustworthiness and reliability determinations that allow individuals to have unescorted access to category 1 or category 2 quantities of radioactive materials possessed by the licensee.

§ 37.23(b)(2)

Each licensee shall name one or more individuals to be reviewing officials. After completing the background investigation on the reviewing official, the licensee shall provide, under oath or affirmation, a certification that the reviewing official is deemed trustworthy and reliable by the licensee. The fingerprints of the named reviewing official must be taken by a law enforcement agency, Federal or State agencies that provide fingerprinting services to the public, or commercial fingerprinting services authorized by a State to take fingerprints. The licensee shall recertify that the reviewing official is deemed trustworthy and reliable every 10 years in accordance with 10 CFR 37.25(b).

Explanation

Only ROs may make T&R determinations authorizing individuals to have unescorted access. ROs must themselves undergo background investigations, and the licensee must certify by oath or affirmation that the RO is trustworthy and reliable.

Q&As

Q1: What is the role of the RO?

A1: The RO makes the T&R determinations for the licensee and determines who may be granted unescorted access authorization. Note that the IC fingerprinting orders refer to a T&R official as the individual who made determinations on an employee's T&R. Unlike the RO, the T&R official did not have to be fingerprinted and subject to a background investigation. This rule closes that potential security gap by requiring a T&R determination for each RO.

Q2: Who can be a reviewing official?

A2: The licensee decides who is named as an RO. The RO may be the radiation safety officer (RSO), someone from the Human Resources (HR) department, or any other individual. To review the FBI criminal history records, the RO must be a licensee employee, not a contractor. Also, the RO must be authorized for unescorted access to category 1 or category 2 quantities of radioactive material or access to SGI or SGI-M, but the RO does not have to be actually granted access to the material.

Q3: May the HR department be designated to perform the background investigations and be the repository for T&R determination records? If we have a process in place, may we continue to use that process? Does the RSO have to be involved?

A3: The review or record storage or both may be delegated to a licensee's HR department or any other appropriate department, depending on its mission and available resources. Additionally, a licensee may use the information previously obtained through the hiring process or another background investigation process to support a T&R determination without having to reverify the information. However, the individual responsible for T&R determinations must also undergo fingerprinting and an FBI criminal history check, and the licensee must document the basis for concluding that its ROs are trustworthy and reliable.

The RSO does not have to be involved in T&R determinations. Because safety and security often go hand in hand, however, the RSO likely will need to be integrally involved in any implementation decisions.

Q4: Must individuals whose sole job duties are to determine eligibility for employment (e.g., HR specialists) undergo a background investigation and be determined to be trustworthy and reliable?

A4: No. Only the RO must undergo a background investigation and must be determined trustworthy and reliable because an untrustworthy RO could approve access by other untrustworthy individuals as part of a conspiracy to steal, sabotage, or divert radioactive material. However, the T&R determination may use relevant information obtained and evaluated by HR personnel when determining whether to hire an individual. If the T&R determination is not made by the HR staff and if the licensee relies only on the HR staff to obtain information from the background check (e.g., employment history, education, and personal references), the HR staff does not have to undergo a background investigation.

Q5: How do I name an individual to be a reviewing official?

A5: The licensee needs to submit the individual's name and fingerprints to the NRC. Either a law enforcement agency, a Federal or State agency that provides fingerprinting services to the public, or a commercial fingerprinting service authorized by a State to take fingerprints need to be the one to take the fingerprints of the individual(s) under consideration for an RO. After completing the background investigation on the candidate RO and upon determining that he or she is trustworthy and reliable, the licensee must provide, under oath or affirmation, a certification that it deems the RO to be trustworthy and reliable. In accordance with 10 CFR 37.25(b), the licensee must also recertify that the RO is trustworthy and reliable every 10 years after his or her selection. Annex A, "Additional Guidance for Evaluating an Individual's Trustworthiness and Reliability for Allowing Unescorted Access to Certain Radioactive Material," at the end of the Q&As for this subpart provides additional guidance for evaluating an individual's T&R for allowing unescorted access.

Q6: If a licensee already has a T&R official or RO to comply with an NRC security order, may this individual continue to serve as the RO under 10 CFR Part 37?

A6: If the T&R official or RO required by an NRC security order has undergone fingerprinting and a background investigation, he or she may continue to serve as the RO without further action. If the individual serving as the T&R official has not undergone fingerprinting and a background investigation, the licensee must complete these steps before he or she can make any additional T&R determinations.

Q7: Section 10 CFR 37.23(b)(2) requires that the fingerprints of the RO be taken by a law enforcement agency, a Federal or State agency that provides fingerprinting services to the public, or a commercial fingerprinting service authorized by a State to take fingerprints. This makes the licensee incur additional cost to travel to an authorized agency and fees to have the authorized agency take fingerprints. Why doesn't the rule allow the RO to be fingerprinted by the licensee's personnel?

A7: Because the RO has greater responsibility in the access authorization program and because he or she will be making the determinations to allow access, the NRC has determined that an independent entity must take the RO's fingerprints to ensure that the identification provided matches the person being fingerprinted. This requirement ensures the correct identification of the individual submitting the fingerprints. Without this requirement, the RO could substitute as his or her own fingerprints those of another individual who is known not to have a criminal history or terrorist ties.

Q8: How do I know that a new applicant for RO duties isn't someone who's been denied unescorted access in some other State and has just come to me to get approval? Will the NRC or some other organization keep a registry of such applicants?

A8: If the licensee does an adequate employment history check and learns that the applicant worked for other licensees, it may be able to forestall this problem by specifically asking one or more of his or her most recent employers. If none of these previous employers are willing to provide information on previous denials of access authorization, the licensee may try to get information from friends, relatives, or other sources independent of the applicant's references under 10 CFR 37.25(a)(6). The licensee also may ask the applicant in a personal interview and look for nonverbal cues that may indicate that he or she is lying or withholding relevant information. The NRC does not have a system to track individuals that have been granted unescorted access or to track those that have been denied.

Q9: What if a licensee's reinvestigation of its RO, or new information from any source at any time between reinvestigations, requires the licensee to revoke the RO's access authorization? Do all of his or her T&R determinations on other individuals since the previous investigation need to be redone?

A9: The answer to this question will depend on the kind of new information that the licensee obtains about the RO. For example, a discovery that the individual is on a no-fly list for suspected collaboration with a known criminal or terrorist organization should warrant a review of all his or her past T&R determinations or at least those dating from about the time that he or she was first put on such a list. The licensee is responsible for ensuring that only trustworthy and reliable individuals are authorized to have access to radioactive materials. The licensee's assessment of the extent of its potential security liability should include, but may not be limited to, such factors as the number of employees to whom the RO has authorized unescorted access since his or her last background check; the licensee's knowledge of the kinds and frequency of social, financial, familial, or romantic contact that the RO has had with any of these employees; and the timing of their access authorization approvals in relation to the timing of the event(s) that raised doubts about the former RO's T&R.

§ 37.23, "Access Authorization Program Requirements" (continued)

§ 37.23(b), "Reviewing Officials" (continued)

§ 37.23(b)(3)

Reviewing officials must be permitted to have unescorted access to category 1 or category 2 quantities of radioactive materials or access to safeguards information-modified handling if the licensee possesses safeguards information or safeguards information-modified handling.

§ 37.23(b)(4)
Reviewing officials cannot approve other individuals to act as reviewing officials.

§ 37.23(b)(5)
A reviewing official does not need to undergo a new background investigation before being named by the licensee as the reviewing official if:

§ 37.23(b)(5)(i)
The individual has undergone a background investigation that included fingerprinting and a FBI criminal history records check and has been determined to be trustworthy and reliable by the licensee; or

§ 37.23(b)(5)(ii)
The individual is subject to a category listed in 10 CFR 37.29(a).

Explanation

ROs must be permitted unescorted access to category 1 or category 2 radioactive materials or to SGI or SGI-M if the licensee possesses this information. ROs may not approve other individuals to act as ROs. An RO does not need to undergo a new background investigation if he or she has already been determined trustworthy and reliable after a background investigation that included fingerprinting and a criminal history records check.

Q&As

Q1: When may an RO make T&R determinations to permit unescorted access by employees?

A1: The RO may make T&R determinations for any employee who requires unescorted access only after the licensee has completed the required fingerprinting, FBI criminal history records check, and background investigation of the RO and has certified to the NRC, under oath or affirmation, that she or he is trustworthy and reliable.

Q2: May I appoint multiple ROs?

A2: Yes. A licensee may appoint multiple ROs; however, each RO must undergo fingerprinting, a background investigation, and an FBI criminal history records check, and the licensee must certify, under oath or affirmation, that each RO is trustworthy and reliable. To review FBI criminal history records, each RO must be a licensee employee, not a contractor.

Q3: 10 CFR 37.23(b)(4) prohibits a reviewing official from approving other individuals to act as reviewing officials. May the incumbent RO do anything related to a licensee's nomination of another reviewing official, or is the incumbent barred from providing any support?

A3: An approved RO may conduct the other aspects of the background investigation 10 CFR 37.25 (a)(2) through 10 CFR 37.25(a)(7)) for an individual whom the licensee will nominate as an RO.

Q4: May a State official exempted under 10 CFR 37.29(a)(6) or 10 CFR 37.29(a)(7) serve as a reviewing official?

A4: Yes. An official qualifying for relief from fingerprinting and other background investigation requirements under this subsection may serve as an RO, as would an individual covered by any of the other categories of relief. However, such a scenario is not likely to occur unless the official serves a State entity licensed by its Agreement State regulatory agency.

Q5: Why must ROs be permitted unescorted access to category 1 or category 2 quantities of radioactive materials or access to SGI?

A5: It is important that individuals making the final determination of T&R for others be trustworthy and reliable themselves. This means that ROs must undergo the same background investigation as individuals granted unescorted access, including fingerprinting and the FBI criminal records check. If the RO is not fingerprinted, a potentially exploitable vulnerability could be created in the security program. The NRC's AEA authority to collect fingerprints only applies to individuals who have unescorted access to radioactive material or access to SGI. The RO must therefore be permitted access to radioactive material or SGI to give the NRC the authority to require the collection of his or her fingerprints for submittal to the FBI for processing. However, the RO will not need to actually use the unescorted access if the licensee determines that the RO has no need for it to perform a job duty.

§ 37.23, "Access Authorization Program Requirements" (continued)

§ 37.23(c), "Informed Consent"

§ 37.23(c)(1)

Licensees may not initiate a background investigation without the informed and signed consent of the subject individual. This consent must include authorization to share personal information with other individuals or organizations as necessary to complete the background investigation. Before a final adverse determination, the licensee shall provide the individual with an opportunity to correct any inaccurate or incomplete information that is developed during the background investigation. Licensees do not need to obtain signed consent from those individuals that meet the requirements of § 37.25(b). A signed consent must be obtained prior to any reinvestigation.

§ 37.23(c)(2)

The subject individual may withdraw his or her consent at any time. Licensees shall inform the individual that:

§ 37.23(c)(2)(i)

If an individual withdraws his or her consent, the licensee may not initiate any elements of the background investigation that were not in progress at the time the individual withdrew his or her consent; and

§ 37.23(c)(2)(ii)

The withdrawal of consent for the background investigation is sufficient cause for denial or termination of unescorted access authorization.

Explanation

Licensees may not begin a background investigation without the informed and signed consent of the subject individual. The subsection also requires a licensee to inform the subject individual of his or her right to correct inaccurate or incomplete information and to have an opportunity to correct it before the licensee makes an adverse T&R determination. If an individual withdraws consent to a background investigation, the licensee may not begin any new element of that investigation.

Q&As

Q1: What is informed consent?

A1: Informed consent is the individual's authorization that allows the licensee to conduct the background investigation to determine if the individual is trustworthy and reliable. The licensee needs to explain to the individual that a background investigation is being conducted and then explain the potential consequences if the individual does not agree to the background investigation. The signed consent shows that the individual has been provided the appropriate explanation and indicates his or her understanding that a background investigation will be conducted. The signed consent must include authorization to share personal information with other individuals or organizations, as necessary, to complete the background investigation. Annex B, "Sample Consent Form for Background Investigations," to Subpart B provides a template for a possible consent form that a licensee can adapt for its use.

Q2: For individuals who have already been granted unescorted access under the various orders, does a licensee need to go back and obtain informed consent?

A2: No. The licensee does not need to go back and obtain informed consent from these individuals unless they have not already undergone a background investigation that included fingerprinting and an FBI criminal history records check. The licensee must obtain an informed consent before starting any new background investigation and before a reinvestigation of any currently approved individual.

Q3: How do I obtain informed consent from an individual?

A3: The licensee should first explain the informed consent process to the individual either orally or in writing. The easiest way to obtain the informed consent is to have a form with the necessary information that the individual could read and then sign and date. Annex B provides a template for a possible consent form.

Q4: What should I do if an individual withdraws his or her consent and I'm in the middle of his or her background investigation?

A4: If an individual withdraws his or her consent, 10 CFR 37.23(c)(2)(ii) requires the licensee to inform the individual that withdrawal of consent for the background investigation is sufficient cause for denial or termination of unescorted access authorization. The licensee may finish work on any elements of the background investigation that it had undertaken before the consent was withdrawn; however, the licensee cannot, under 10 CFR 37.23(c)(2)(i), initiate any additional elements that were not already in process.

Q5: If an individual withdraws his or her consent and I terminate the background investigation, may the individual be granted unescorted access to the material?

A5: No. If an individual withdraws consent for the background investigation and if that investigation is therefore never completed, the licensee would not have a basis for granting that individual unescorted access.

Q6: May an individual who initially withdraws permission for a background investigation, but later gives permission, be granted unescorted access?

A6: Yes. If the completed background investigation supports the determination, the individual may be granted unescorted access. The fact that permission was initially withdrawn would not, in itself, be sufficient grounds for denial.

§ 37.23, "Access Authorization Program Requirements" (continued)

§ 37.23(d), "Personal History Disclosure"

Any individual who is applying for unescorted access authorization shall disclose the personal history information that is required by the licensee's access authorization program for the reviewing official to make a determination of the individual's trustworthiness and reliability. Refusal to provide, or the falsification of, any personal history information required by this subpart is sufficient cause for denial or termination of unescorted access.

Explanation

These provisions establish that individuals applying for unescorted access authorization must disclose their personal history information.

Q&As

Q1: What is a personal history disclosure?

A1: The personal history disclosure is the information that the individual who is seeking unescorted access to category 1 or category 2 quantities of radioactive material must provide. It is the type of information typically collected on an employment application. The information should include items, such as employment history, education, references, and any arrest record. It may also include, but is not required to include, information related to finances, such as bankruptcies. This information provides the RO with a starting point for the background investigation.

Q2: The information sounds like information provided for employment. May I use an employment application to gather the information?

A2: The information provided under a personal history disclosure is similar to information obtained by many companies in an application for employment. If the employment application contains adequate information, the licensee may use it for this purpose.

§ 37.23, "Access Authorization Program Requirements" (continued)

§ 37.23(e), "Determination Basis"

§ 37.23(e)(1)

The reviewing official shall determine whether to permit, deny, unfavorably terminate, maintain, or administratively withdraw an individual's unescorted access authorization based on an evaluation of all of the information collected to meet the requirements of this subpart.

§ 37.23(e)(2)

The reviewing official may not permit any individual to have unescorted access until the reviewing official has evaluated all of the information collected to meet the requirements of this subpart and determined that the individual is trustworthy and reliable. The reviewing official may deny unescorted access to any individual based on information obtained at any time during the background investigation.

§ 37.23(e)(3)

The licensee shall document the basis for concluding whether or not there is reasonable assurance that an individual is trustworthy and reliable.

§ 37.23(e)(4)

The reviewing official may terminate or administratively withdraw an individual's unescorted access authorization based on information obtained after the background investigation has been completed and the individual granted unescorted access authorization.

§ 37.23(e)(5)

Licensees shall maintain a list of persons currently approved for unescorted access authorization. When a licensee determines that a person no longer requires unescorted access or meets the access authorization requirement, the licensee shall remove the person from the approved list as soon as possible, but no later than 7 working days, and take prompt measures to ensure that the individual is unable to have unescorted access to the material.

Explanation

These provisions establish requirements for ROs to grant, deny, or withdraw an individual's unescorted access authorization. Licensees must document the reason that an individual is determined to be trustworthy and reliable and must maintain a list of individuals approved for unescorted access.

Q&As

Q1: What information should the reviewing official use to determine that an individual is trustworthy and reliable?

A1: The regulation in 10 CFR 37.23(e)(2) requires the RO to use all the information gathered during the background investigation to make a determination that an individual is trustworthy and reliable. The NRC expects licensees to use their best efforts to obtain this information. The licensee may use information previously obtained during the hiring process to support the T&R determination without having to reverify that information.

The RO may deny unescorted access to any individual based on any information obtained at any time during the background investigation that calls into question the individual's T&R. However, the licensee may not, under 10 CFR 37.27(b), base a final determination to deny an individual unescorted access to category 1 or category 2 quantities of radioactive material solely on the basis of information received from the FBI involving (1) an arrest of more than 1 year old for which no information is available on the disposition of the case or (2) an arrest that resulted in dismissal of the charge or an acquittal. A record on the disposition of the case may not exist because information on a dismissal or acquittal may not have been recorded. Annex A lists some indicators that licensees should consider as potential T&R concerns.

Q2: How will we obtain all the information we need to complete a background investigation before granting unescorted access to an individual without a previous employment record (e.g., a recent high school or college graduate)?

A2: For new hires without an employment history, the licensee will need to rely more on other aspects of the background investigation, such as references. A lack of employment history does not need to be a negative consideration in determining if an individual is deemed to be trustworthy and reliable and is given unescorted access to category 1 or category 2 quantities of radioactive material. The individual may be escorted until the licensee completes the background investigation and has made a determination on the individual's T&R for unescorted access.

Q3: Are managers subject to the same background investigations?

A3: Yes. If a manager has unescorted access to the radioactive material, he or she is subject to the same background investigation requirements.

Q4: What criteria do I use to determine trustworthiness and reliability?

A4: The NRC has not developed a set of criteria for determining T&R because no such list is likely to cover all licensees' needs, and each licensee is in the best position to weigh the many considerations that must support such determinations. Therefore, the licensee is responsible for making T&R determinations for all employees granted unescorted access. The background investigations under 10 CFR 37.25 are designed to identify past actions that might call into question an individual's T&R. Annex A lists some indicators that licensees should consider as potential concerns. Although the licensee should review this annex in its entirety, the following indicators are provided for convenience:

- impaired performance attributable to psychological or other disorders
- conduct that warrants referral for criminal investigation or that results in an arrest or a conviction
- indication of deceitful or delinquent behavior
- attempted or threatened destruction of property or life

- suicidal tendencies or attempted suicide
- illegal drug use or the abuse of legal drugs
- alcohol abuse disorders
- recurring financial irresponsibility
- irresponsibility in the performance of assigned duties
- inability to deal with stress or the appearance of being under unusual stress
- failure to comply with work directives
- hostility or aggression toward fellow workers or authority
- uncontrolled anger, violation of safety or security procedures, or repeated absenteeism
- significant behavioral changes, moodiness, or depression

However, these indicators are neither meant to be all inclusive nor intended to be disqualifying factors. Licensees also may consider extenuating or mitigating factors in their determinations.

Q5: Are you requiring licensees to determine if employees are telling the truth?

A5: No. The regulation in 10 CFR 37.23(e)(1) only requires licensees to evaluate all the information collected to meet the requirements of this subpart and to document, under 10 CFR 37.23(e)(3), the basis for concluding whether reasonable assurance that an individual is trustworthy and reliable exists.

Q6: If someone has a drug or alcohol problem, does that automatically make him or her unreliable?

A6: No. A drug or alcohol problem does not, in itself, make an individual unreliable, especially if the individual has acknowledged the problem and is undergoing treatment.

Q7: How should the T&R determination for my employees be documented?

A7: The regulation in 10 CFR 37.23(e)(3) requires that the documentation include the individual employee's or applicant's name and the basis for concluding if reasonable assurance exists that he or she is trustworthy and reliable. The elements that the licensee is required to consider in making its determination appear in 10 CFR 37.25. Because the basis for the licensee's determination is also the result of a process, a good documentation practice would include the criteria, procedures, and records that the licensee used to support its determination; the date of this determination; and the name and signature of the person responsible for making it. In cases for which the licensee has been unable to obtain adequate information on an element, such as a character and reputation determination, it should make this clear and should describe the efforts it made to gather the needed information.

The type of documentation may include a facsimile, voice mail, e-mail records, letters, photographs, film, audio or video tape, notes to a file, or anything else that the licensee used to support its determination. The form of the documentation may be an original or copy, may be on paper, may be electronic, or may be other types of media as long as the record is legible for the required period and can be accessed. To discourage unauthorized alterations, electronic copies should, if possible, be read-only, such as those in portable document format (PDF).

Q8: If I can't obtain all of the information required in 10 CFR 37.25, may I still grant the individual unescorted access?

A8: Yes. If the licensee concludes that the individual should still be authorized for unescorted access based on other background investigation information, it may grant the individual unescorted access. The regulation in 10 CFR 37.23(e) requires the licensee to document the basis for its decision to grant unescorted access. If a previous employer, educational institution, or any other entity with which the individual claims to have been engaged fails to provide information at least 10 business days after the licensee's request, if such entities indicate an inability or unwillingness to provide information, or if the licensee is unable to reach the entity, the regulation in 10 CFR 37.25(a)(7) requires the licensee to document that and to attempt to obtain the information from an alternate source.

The licensee also has the option of escorting the individual and not making a T&R determination.

Q9: If I've determined someone to be trustworthy and reliable, and that individual later takes the material for malevolent use, what actions are expected of me? What liability do I assume because of my T&R determination?

A9: If nothing in the background investigation caused a licensee to deny access and if the licensee did everything that was required, it would not be in violation of the access authorization requirements. The licensee must provide reasonable assurance that persons granted access are trustworthy and reliable. If the licensee fails to provide that assurance, it will be in violation of the 10 CFR Part 37 requirements, and the NRC will consider enforcement action. However, providing assurance means that the licensee has made a reasonable effort, as required by 10 CFR Part 37, to ascertain T&R and has documented its actions. As long as the licensee can show that it has made a reasonable good-faith effort, the NRC will not second-guess its decision.

Q10: Does the denial of unescorted access create legal liability for the licensee?

A10: Although the licensee should check the requirements of applicable State laws, the NRC does not consider a denial of unescorted access authorization to be a denial of employment. The applicant may still work in areas of the facility outside of security zones or may perform escorted work within the facility. A denial only prevents the employee from having unescorted access to a category 1 or category 2 quantity of material.

Q11: How can we address the unique challenges related to establishing T&R for foreign nationals?

A11: Determining the T&R of foreign nationals, including students, does pose special challenges. An evaluation of academic and other references (e.g., transcripts, college applications, and financial aid applications) may form part of the basis for a T&R determination. A visa, in and of itself, does not provide an adequate basis for determining that the individual is trustworthy and reliable.

Background investigations are required to verify and develop information that supports the basis for the T&R determination. The regulation in 10 CFR 37.25(a)(6) requires licensees to obtain independent corroborating information to the extent possible. The 10 CFR Part 37 requirements incorporate the phrase "to the extent possible" to communicate the expectation that licensees will use their best effort to obtain the information required. However, if obtaining such corroborative information becomes impossible, and if the licensee concludes that the individual should still be authorized for unescorted access based on other background investigation information, it should be prepared to document its efforts to obtain the necessary information.

Q12: If I decide that one of my employees previously granted unescorted access to radioactive material should no longer have unescorted access, what actions can I take?

A12: The licensee should immediately revoke the individual's unescorted access because it is ultimately responsible for determining the best method to revoke an employee's unescorted access to radioactive material. The licensee should document any change in access status with supporting information.

Q13: Does my list of individuals who have been granted unescorted access need to include all employees and contractors who have been denied unescorted access?

A13: No. The list only needs to include those individuals to whom the licensee has decided to grant unescorted access. The list is not intended to include unapproved individuals who must be escorted.

Q14: Are there any other sources of information I should check before making a final determination on an individual?

A14: Yes. The licensee should check the NRC's list of escalated enforcement actions issued to individuals. The list includes individuals who are prohibited from working with radioactive materials. This list appears on the NRC's Web site at http://www.nrc.gov/reading-rm/doccollections/enforcement/actions/individuals/.

Q15: How quickly must I remove an individual from my list of those approved for unescorted access after I've determined that he or she no longer requires it?

A15: The regulation in 10 CFR 37.23(e)(5) requires the licensee to remove such an individual from its approved list as soon as possible but not later than 7 working days after it has determined that he or she no longer requires unescorted access to the radioactive material subject to this part. In addition, the regulation in 10 CFR 37.23(e)(5) requires the licensee to take prompt measures to ensure that he or she is no longer able to have unescorted access.

§ 37.23, "Access Authorization Program Requirements" (continued)

§ 37.23(f), "Procedures"

Licensees shall develop, implement, and maintain written procedures for implementing the access authorization program. The procedures must include provisions for the notification of individuals who are denied unescorted access. The procedures must include provisions for the review, at the request of the affected individual, of a denial or termination of unescorted access authorization. The procedures must contain a provision to ensure that the individual is informed of the grounds for the denial or termination of unescorted access authorization and allow the individual an opportunity to provide additional relevant information.

Explanation

These provisions establish requirements for written procedures for implementing the access authorization program.

Q&As

Q1: Am I as a licensee required to have procedures for conducting background investigations?

A1: Although 10 CFR 37.23(f) does not specifically require procedures for conducting background investigations, it does require the licensee to develop, implement, and maintain written procedures for implementing the access authorization program as a whole. The NRC expects that the licensee would have procedures for conducting background investigations as an important component of the procedures for its access authorization program. For example, the regulations in 10 CFR 37.23(f) require that the licensee's procedures enable applicants to review the outcome of its T&R determination, and under 10 CFR 37.23(e), the licensee's determination must be based on information produced from a background investigation. The licensee's procedures under this subsection must also address the notification of individuals denied authorization for unescorted access, must ensure that these individuals are informed of the grounds for the licensee's denial or termination of unescorted access authorization, and must provide such individuals an opportunity upon request to review the licensee's denial or termination and additional relevant information. Thus, the licensee should, as good practice, ensure that its procedures address its expectations for the performance and documentation of background investigations.

Q2: What procedures would you suggest for making T&R determinations?

A2: The NRC does not suggest specific procedures for making T&R determinations.

However, licensees should generally consider developing procedures that address, among other things, how to conduct a background investigation; how to develop and document a determination basis, the criteria for when the licensee would accept another company's findings, and what information the licensee needs to verify; how to reinstate an individual; how to maintain the list of individuals approved for unescorted access; and how to withdraw an individual's approved unescorted access. Procedures may differ in kind, sequence, structure, and level of detail as long as they produce the kinds of information and documentation required by 10 CFR 37.25 and 10 CFR 37.27.

§ 37.23, "Access Authorization Program Requirements" (continued)

§ 37.23(g), "Right To Correct and Complete Information"

§ 37.23(g)(1)

Prior to any final adverse determination, licensees shall provide each individual subject to this subpart with the right to complete, correct, and explain information obtained as a result of the licensee's background investigation. Confirmation of receipt by the individual of this notification must be maintained by the licensee for a period of 1 year from the date of the notification.

§ 37.23(g)(2)

If after reviewing his or her criminal history record an individual believes that it is incorrect or incomplete in any respect and wishes to change, correct, update, or explain anything in the record, the individual may initiate challenge procedures.

These procedures include direct application by the individual challenging the record to the law enforcement agency that contributed the questioned information or a direct challenge as to the accuracy or completeness of any entry on the criminal history record to the Federal Bureau of Investigation, Criminal Justice Information Services (CJIS) Division, ATTN: SCU, Mod. D-2, 1000 Custer Hollow Road, Clarksburg, WV 26306 as set forth in 28 CFR 16.30 through 16.34. In the latter case, the Federal Bureau of Investigation (FBI) will forward the challenge to the agency that submitted the data and will request that the agency verify or correct the challenged entry. Upon receipt of an official communication directly from the agency that contributed the original information, the FBI Identification Division makes any changes necessary in accordance with the information supplied by that agency. Licensees must provide at least 10 days for an individual to initiate action to challenge the results of an FBI criminal history records check after the record being made available for his or her review. The licensee may make a final adverse determination based upon the criminal history records only after receipt of the FBI's confirmation or correction of the record.

Explanation

These provisions require the licensee to give individuals the right to complete, correct, and explain the information gathered for the T&R determination before it makes an adverse determination. The licensee must also allow these individuals the right to challenge that information before it makes an adverse T&R determination.

Q&As

Q1: Why must I provide an applicant for unescorted access authorization the right to "complete, correct, and explain" the information I obtained during a background investigation before I make a final adverse determination (i.e., deny unescorted access)?

A1: An individual is given the right to review the record before any adverse determination because the information that the licensee is relying on to make that determination must be complete, correct, and fully explained. The licensee does not want to make an adverse decision based on possibly incomplete or inaccurate information. Misinformation could be the result of a recordkeeping error, identity theft, or misidentification of another individual with the same or similar name. An individual also may have been charged with an offense of some type and later have been found innocent or had the charges dropped, and the record might never have been updated to reflect this additional information. The individual might also be able to provide mitigating information that could be relevant to the licensee's decision.

Q2: May any of the background investigation information be challenged?

A2: Yes. The individual may request clarification of any information that he or she believes is incorrect. For some kinds of information, the individual may simply provide the licensee with what he or she believes is the correct information. The licensee may then consider the new information in its T&R determination process. However, the licensee should verify the corrected information, if possible.

Q3: What is the process for challenging criminal history records?

A3: If the individual believes that his or her criminal history records are incorrect or incomplete in any respect, he or she may initiate challenge procedures. These procedures would include direct application by the individual to the law enforcement agency that contributed the questioned information. An individual may also challenge the accuracy or completeness of any entry on the criminal history record by contacting the FBI at Federal Bureau of Investigation, Criminal Justice Information Services (CJIS) Division, ATTN: SCU, Mod. D-2, 1000 Custer Hollow Road, Clarksburg, WV 26306.

Instructions for a challenge to the FBI appear in 28 CFR 16.30, "Purpose and Scope." The FBI will forward the challenge to the agency that submitted the data and will request that the agency verify or correct the challenged entry. Upon receipt of an official communication directly from the agency that contributed the original information, the FBI Identification Division will make any necessary changes to the individual's criminal history record.

The licensee may not make a final adverse determination based solely on the criminal history records until the licensee receives the FBI's confirmation or correction of the record.

Q4: How long must I allow for an individual to challenge my findings before I make a final determination to grant or deny authorization for unescorted access?

A4: The regulation in 10 CFR 37.23(g)(2) requires the licensee to allow at least 10 days for an individual to initiate action to challenge the results of an FBI criminal history records check, although it may allow more time. The licensee may use its judgment for other elements of the background investigation. The implementation procedures for the licensee's access authorization program should specify its timeframes for these challenges. In any case, the licensee may not make a final adverse determination based on the criminal history records until it has received the FBI's confirmation or correction of the record(s) that the individual has challenged.

§ 37.23, "Access Authorization Program Requirements" (continued)

§ 37.23(h), "Records"

§ 37.23(h)(1)

The licensee shall retain documentation regarding the trustworthiness and reliability of individual employees for 3 years from the date the individual no longer requires unescorted access to category 1 or category 2 quantities of radioactive material.

§ 37.23(h)(2)

The licensee shall retain a copy of the current access authorization program procedures as a record for 3 years after the procedure is no longer needed. If any portion of the procedure is superseded, the licensee shall retain the superseded material for 3 years after the record is superseded.

§ 37.23(h)(3)

The licensee shall retain the list of persons approved for unescorted access authorization for 3 years after the list is superseded or replaced.

Explanation

These provisions establish recordkeeping requirements for licensee access authorization programs.

Q&As

Q1: What access authorization records am I required to retain?

A1: The NRC considers "documentation regarding the trustworthiness and reliability of individual employees" under 10 CFR 37.23(h)(1) to include all fingerprint and criminal history records received from the FBI, or a copy if the individual's file has been transferred, and all records from the other background investigation requirements under 10 CFR 37.25(a)(2) through 10 CFR 37.25(a)(7). The licensee also is required to retain any written confirmations received from entities concerning a Federal security clearance under 10 CFR 37.29(a)(12), any written confirmations received concerning a favorably adjudicated criminal history records check under 10 CFR 37.29(b), and any written verifications received from service providers under 10 CFR 37.29(a)(13). The licensee must also retain a record of its determination bases (reasons for granting unescorted access or denying it) under 10 CFR 37.23(e)(3). The determination documentation must include the individual's name and should include the criteria, procedures, and supporting documentation that the licensee used in the process of making the determination for each individual); the date of the determination; and the name and signature of the RO responsible for making it. The licensee also is required to retain copies of the implementation procedures. (See also the Q&As on 10 CFR 37.23(e).)

Q2: What form of documentation do I need to keep for the access authorization program? Must I convert my paper documents to a digital format?

A2: No. The licensee may choose to scan paper and other documents into digital format for its own convenience and economy of storage space. This, however, is not a requirement. The form of the documentation may be an original or copy, may be on paper, may be electronic, or may be in other forms of media as long as the record is legible for the required period and can be accessed. To discourage unauthorized alterations, electronic copies should, if possible, be read-only, such as those in PDF.

Q3: What procedures do I need to keep for the access authorization program?

A3: The regulation in 10 CFR 37.23(h)(2) requires the licensee to retain copies of all the procedures necessary to implement the access authorization program, including procedures for obtaining written consent, conducting the investigation and reinvestigation, correcting the record, challenging a determination to deny authorization for unescorted access, and documenting the final determination.

Q4: Do I need to keep any information relating to the lists of individuals approved for unescorted access other than the lists themselves?

A4: No. The regulation in 10 CFR 37.23(h)(3) only requires the licensee to retain a copy of the list of individuals who have current unescorted access to the material. When a list is updated, the previous list must be kept as a record for 3 years. The licensee does not need to retain other list-related records, although it will need to retain background investigation and determination-basis-related records for each individual on the list.

Q5: Must I also keep a list of individuals who have been denied access?

A5: No. The fact that someone is not included on the access list means that they should not be granted unescorted access to the material, and a list of individuals who have been denied access is not necessary.

Q6: How long must the records be maintained?

A6: The licensee must maintain background investigation records for 3 years after the individual no longer requires unescorted access to category 1 or category 2 quantities of radioactive material. The regulation in 10 CFR 37.23(e)(3) requires that these records also include those that provide the basis for the licensee's determination on if an individual is trustworthy and reliable for unescorted access authorization. The licensee must maintain the list of individuals approved for unescorted access authorization for 3 years after the list is superseded or replaced. The regulation in 10 CFR 37.23(h)(3) requires the licensee to retain a procedure for 3 years after the it has been replaced with a new or revised version.

Q7: If a facility closes and the license is terminated, does the licensee need to keep records for an additional 3 years?

A7: No. Once the Commission terminates a license for any reason, the licensee is no longer required under 10 CFR 37.103, "Record Retention," to maintain its records for the access authorization program, and these records may be destroyed. (See also 10 CFR 37.31, "Protection of Information," for requirements on the protection of personal information.)

§ 37.25, "Background Investigations"

§ 37.25(a), "Initial Investigation"

Before allowing an individual unescorted access to category 1 or category 2 quantities of radioactive material or to the devices that contain the material, licensees shall complete a background investigation of the individual seeking unescorted access authorization. The scope of the investigation must encompass at least the 7 years preceding the date of the background investigation or since the individual's eighteenth birthday, whichever is shorter.

Explanation

These provisions establish the requirement to conduct a background investigation for individuals who need unescorted access to category 1 and category 2 quantities of radioactive material.

Q&As

Q1: How far back in time must I look into an individual's historical information as part of the background investigation?

A1: The scope of the investigation must encompass at least the 7 years preceding the date of the background investigation or since the individual's 18th birthday, whichever is shorter. To the extent possible, the licensee should look into an individual's history as far back as necessary to be satisfied that sufficient information is available to meet its criteria for determining T&R.

Q2: What are the components of a background investigation?

A2: A background investigation includes several components, including fingerprinting and an FBI identification and criminal history records check, verifications of the individual's true identity, a review of information obtained about the individual's employment history and education, and a character and reputation determination.

Q3: Why are background investigations and T&R determinations necessary?

A3: The background investigation is a tool to obtain information necessary to make decisions about whether an individual is trustworthy and reliable and may be permitted unescorted access to a category 2 or greater quantity of radioactive material. The licensee must ensure that an individual who is seeking unescorted access to radioactive material is dependable in judgment, character, and performance such that unescorted access by that individual to category 1 or category 2 quantities of radioactive material does not constitute an unreasonable risk to the public health and safety or security.

A T&R determination provides the licensee's decision with reasonable assurance that the individual allowed unescorted access will not use the material for malicious purposes. This requirement goes beyond access control for radiation protection purposes and further limits access to only those individuals who have a legitimate need to access the licensed material or device.

Q4: Are these background investigation requirements equivalent to those used in nuclear power plants?

A4: No. Nuclear power plants have additional requirements, such as a credit history check, a psychological assessment, and a behavioral observation program to determine an employee's T&R. (See 10 CFR 73.56, "Personnel Access Authorization Requirements for Nuclear Power Plants," and 10 CFR 73.57, "Requirements for Criminal History Records Checks of Individuals Granted Unescorted Access to a Nuclear Power Facility or Access to Safeguards Information.")

Q5: How does a background investigation assure trustworthiness and reliability?

A5: No background check can provide total assurance that a person granted unescorted access will not use the material for malicious purposes. The required investigation, however, does provide information to determine with reasonable assurance that the individual is who he or she purports to be. It also provides the licensee with information about if the individual's character, reputation, and behavior might be adverse to the safe and secure operation of its facility. A background investigation can provide the licensee with a reasonable basis to determine that allowing an individual to have unescorted access to the licensed material would not constitute an unreasonable risk of a malevolent use of radioactive materials.

§ 37.25, "Background Investigations" (continued)

§ 37.25(a)(1)
Fingerprinting and an FBI identification and criminal history records check in accordance with 10 CFR 37.27.

Explanation

This provision requires the background investigation to include fingerprinting and an FBI identification and criminal history records check.

Q&As

Q1: Why is the NRC requiring fingerprinting and a criminal history records check for individuals to have unescorted access to category 1 or category 2 quantities of radioactive material?

A1: Section 149 of the AEA requires fingerprinting and an FBI identification and criminal history records check for "any individual who is permitted unescorted access to radioactive materials or other property subject to regulation by the Commission that the Commission determines to be of such significance to the public health and safety or the common defense and security as to warrant fingerprinting and background checks." The Commission determined that category 1 and category 2 quantities of radioactive material are of significance to the public health and safety or the common defense and security. Therefore, individuals who have access to category 1 or category 2 quantities of radioactive material must be fingerprinted.

Fingerprinting an individual for an FBI criminal history records check is an important element of the background investigation. It can provide comprehensive information on an individual's recorded criminal activities within the United States and its territories and on the individual's known affiliations with violent gangs or terrorist organizations. It is one element of the determination of T&R. See the Q&As for 10 CFR 37.27 for more detail.

Q2: Why hasn't the NRC established criteria for making a consistent determination on the finding of the fingerprint reports? Won't this lack of criteria result in inconsistent approval or denial of unescorted access authorizations for people who are trustworthy and reliable?

A2: The NRC took a number of considerations into account when it decided not to provide specific criteria for making decisions about ROs and authorizing unescorted access. Because the individual circumstances of each applicant may vary significantly, each licensee needs the flexibility to establish its own program. In addition, because the information obtained from employment history and other background checks may vary widely and because licensees may not be able to obtain sufficient information to determine compliance with all criteria, the licensee may not be able to avoid having to apply its subjective judgment about an applicant in any case. Moreover, a single set of criteria for application to all cases may not be fair to some individuals who may otherwise be well qualified. For example, a criterion prohibiting the selection of a candidate with any criminal record that is less than 10 years old might disadvantage otherwise well-qualified candidates whom a licensee might prefer after taking the candidate's professional qualifications, employment history, character references, and other more recent information into account. Because the particular circumstances of each individual may vary significantly, each licensee needs the flexibility to establish its own program and use its own criteria.

Although a licensee is allowed to apply criteria that may be inconsistent with those of other licensees, the application of these criteria does not necessarily result in unfairness. One size does not fit all. For example, some licensees may be more comfortable with someone who has a criminal record, whereas other licensees may decide that, with an ample availability of other qualified applicants, an applicant with such a record may be employable but should not be allowed unescorted access to the licensee's radioactive material. A criminal record does not, by itself, mean that an individual is not trustworthy and reliable. The individual may have acknowledged his or her mistake and may now be a model citizen. In addition, a licensee may allow an individual unescorted access, but it may implement additional measures to oversee that individual.

The licensee is in the best position to decide what best fits its needs. Because each licensee will need to decide whether to trust an individual with its assets, it has the greatest stake in making a well-informed decision about the person best qualified to fill such a sensitive position. Thus, licensees should be allowed to develop their own criteria for these decisions.

Q3: May I use a contractor to process FBI criminal history record information to meet this requirement?

A3: No. In response to an NRC request for clarification concerning the use of contractors or other third parties for processing criminal history record information, the FBI analyzed the Commission's AEA authority in 1997 and concluded that the use of private contractors by the NRC and its licensees to receive and process FBI criminal history record information is prohibited.

§ 37.25, "Background Investigations" (continued)

§ 37.25(a)(2), "Verification of True Identity"

Licensees shall verify the true identity of the individual who is applying for unescorted access authorization to ensure that the applicant is who he or she claims to be. A licensee shall review official identification documents (e.g., driver's license; passport; government identification; certificate of birth issued by the state, province, or country of birth) and compare the documents to personal information data provided by the individual to identify any discrepancy in the information. Licensees shall document the type, expiration, and identification number of the identification document or maintain a photocopy of identifying documents on file in accordance with § 37.31. Licensees shall certify in writing that the identification was properly reviewed and shall maintain the certification and all related documents for review upon inspection.

Explanation

This provision requires background investigations to include verification of the individual's true identity.

Q&As

Q1: How does a licensee verify true identity?

A1: To verify the identity of an applicant for access authorization under this subsection, the employer must examine "official identification documents" to determine if they reasonably appear to be genuine and if they relate to the individual. The employer should compare the documents to information provided by the applicant. The employer should maintain document information and the certification of review as a record. The employer is not required to determine that the identification is authentic.

Q2: What documents may be used to verify identity?

A2: The licensee may use identity documents issued by a State or local government or by the Federal Government as long as they contain a photograph and information, such as name, date of birth, gender, height, eye color, and address. These documents include passports, drivers' licenses, and identification cards issued by Government entities. The licensee may use one or more of the documentation types required by the U.S. Citizenship and Immigration Service's I-9 form, which applicants use to apply for eligibility for employment. The I-9 form contains a list of acceptable documentation types and is available at no cost at http://www.uscis.gov/portal/site/uscis/menuitem.5af9bb95919f35e66f614176543f6d1a/?vgnextoid=31b3ab0a43b5d010VgnVCM10000048f3d6a1RCRD&vgnextchannel=db029c7755cb9010VgnVCM10000045f3d6a1RCRD. For the complete list of allowable documents, see page 5 of the I-9 form. Note that items 10–12 in List B generally are not acceptable unless the licensee believes that it needs to authorize unescorted access for a minor. These types of documentation (e.g., nursery school records, report cards, and hospital records) are available only to an applicant under the age of 18 using the I-9 form who is unable to provide one of the other documents listed on this form.

§ 37.25, "Background Investigations" (continued)

§ 37.25(a)(3), "Employment History Verification"

Licensees shall complete an employment history verification, including military history. Licensees shall verify the individual's employment with each previous employer for the most recent 7 years before the date of application;

§ 37.25(a)(4), "Verification of Education"

Licensees shall verify that the individual participated in the education process during the claimed period.

Explanation

These provisions require verification of employment history and education as part of the background investigation.

Q&As

Q1: Is the NRC defining 7 years of employment as "uninterrupted" service, or may there be breaks in service?

A1: There may be breaks in service. Under this subsection, however, a licensee must go back a minimum of 7 years unless the individual is younger than 25. (For an individual younger than 25, the licensee only needs to go back to the individual's 18th birthday.) If the individual has gaps in his or her employment record, the licensee should attempt to determine why these gaps exist.

Q2: What kind of employment evaluation needs to be conducted if the employee has been with the licensee 7 years or more? Does the education verification need to be performed for a long-term employee?

A2: The licensee may use its own records of employment for an individual employed with the company more than 7 years; it does not need to check any previous employers. In addition, the licensee does not need to verify an individual's education if such verification occurred more than 7 years ago. If the licensee performed the education verification as a part of the hiring process, it does not need to repeat this check.

Q3: What if a former employer refuses to provide the information I request?

A3: For interviews with past employers, the NRC understands that simple verbal confirmations of past employment and timeframe may be all the information a former employer is willing to provide on an individual. Although a simple confirmation of the nature and timeframe of past employment would not, by itself, enable a licensee to find an individual trustworthy and reliable, it would constitute independent corroboration of the accuracy of the individual's information about that period of his or her employment.

§ 37.25, "Background Investigations" (continued)

§ 37.25(a)(5), "Character and Reputation Determination"

Licensees shall complete reference checks to determine the character and reputation of the individual who has applied for unescorted access authorization. Unless other references are not available, reference checks may not be conducted with any person who is known to be a close member of the individual's family, including, but not limited to, the individual's spouse, parents, siblings, or children, or any individual who resides in the individual's permanent household. Reference checks under this subpart must be limited to whether the individual has been and continues to be trustworthy and reliable.

Explanation

This provision requires the background investigation to include a determination of an applicant's character and reputation.

Q&As

Q1: What constitutes a character and reputation determination?

A1: This is similar to a reference check for employment. The following questions provide examples that the licensee should consider asking when conducting the reference check:

- Would the organization rehire the individual?
- Would it trust the individual with company assets?
- Does it consider the individual to be trustworthy and reliable?
- Has it ever witnessed anything in the individual's behavior that would cause it to question his or her reliability?

A licensee may take into account a number of different kinds of information from a number of sources to make a determination about an individual's character and reputation so long as the information is clearly pertinent to the individual's likely conduct or behavior if he or she were granted unescorted access to any quantity of radioactive material subject to this part. In addition to records of any arrest or conviction as an adult or juvenile felon, examples of considerations pertinent to an individual's T&R may include, but need not be limited to, evidence of false or deceitful statements; loss of a license to drive; repeated high-speed traffic or other violations indicating a reckless disregard for the safety or security of others; a recent bankruptcy, foreclosure, repossession, or garnishment of income; repeated nonpayment of alimony, child support, or lawfully incurred financial obligations for periods of months; repeated instances of personal harassment; or conduct or behavior that would violate any of the licensee's corporate or professional code of ethics or workplace conduct.

Q2: What criteria do I use to evaluate an applicant's character and reputation?

A2: The NRC has not developed a set of criteria for evaluating character and reputation because no such list is likely to cover all individuals' circumstances or all licensees' needs. Each licensee is in the best position to weigh the many considerations that must support such evaluations. In addition to the considerations listed in A1 above, Annex A includes some indicators of character and reputation that licensees should consider for what may be T&R concerns.

Q3: Reference checks under this subpart "must be limited to whether the individual has been and continues to be trustworthy and reliable." What kinds of information are *not* to be considered relevant to an individual's trustworthiness or reliability?

A3: The "not relevant" category includes information about ethnicity; religious affiliation; ideology or political affiliation; sexual orientation; or membership in any organization that does not advocate, perpetrate, or otherwise support violence against persons, damage to property, or criminal activities, including hate crimes. Otherwise, the licensee may consider information about any action, behavior, or conduct that is dishonest, deceitful, a conflict of interest or otherwise unethical, or a violation of any law relevant to an individual's T&R.

Q4: This subsection provides that unless other references are not available, reference checks "may not be conducted with any person who is known to be a close member of the individual's family, including, but not limited to, the individual's spouse, parents, siblings, or children, or any individual who resides in the individual's permanent household." Does this mean that a licensee may not interview or seek information from any member of a subject individual's family in the course of making a determination about that individual's character or reputation?

A4: No. A licensee may contact family members as a reference check if no other references are available.

§ 37.25, "Background Investigations" (continued)

§ 37.25(a)(6)

The licensee shall also, to the extent possible, obtain independent information to corroborate that provided by the individual (e.g., seek references not supplied by the individual); and

§ 37.25(a)(7)

If a previous employer, educational institution, or any other entity with which the individual claims to have been engaged fails to provide information or indicates an inability or unwillingness to provide information within a timeframe deemed appropriate by the licensee but at least after 10 business days of the request or if the licensee is unable to reach the entity, the licensee shall document the refusal, unwillingness, or inability in the record of investigation and attempt to obtain the information from an alternate source.

Explanation

These provisions require licensees to obtain to the extent possible background investigation information independent from that provided by the individual. Licensees must also document the refusal, unwillingness, or inability of a source to provide information.

Q&As

Q1: What type of information is considered independent information to corroborate that provided by the individual?

A1: The licensee may obtain independent information through interviews with anyone who knows or previously knew the individual, such as teachers; friends; coworkers; neighbors; local members of a church, mosque, synagogue, club, or civic association of which the applicant is a member; or family members. Although information obtained from independent sources may not, by itself, suffice to enable a licensee to find an individual trustworthy and reliable, it would constitute independent corroboration of the accuracy of the individual's information about that period of his or her life. It may also provide information that could support a decision not to authorize unescorted access.

Q2: What should I do if an individual or entity contacted as part of a background investigation refuses to respond?

A2: If a previous employer, educational institution, or any other entity fails to provide information or indicates an inability or unwillingness to provide information in a timely manner, the regulation in 10 CFR 37.25(a)(7) requires the licensee to document the refusal, unwillingness, or inability to respond in the record of investigation. The licensee must then attempt to obtain confirmation of the applicant's employment or education history or other personal associations from at least one alternate source.

Past employers are often hesitant to say anything about a past employee for fear of being held liable. If the licensee receives any input from a former employer or other reference (even if he or she refuses to comment), you should document the conversation and take notes about what the former employer or reference said. If attempts to contact the reference fail after several tries, the licensee should make a note of that as well. A licensee may ask the individual for the name of another coworker or a second-line supervisor who may be willing to provide confirmation of employment.

Q3: Won't documenting my attempts to obtain information from uncooperative references and alternate sources of information under 10 CFR 37.25(a)(7) be excessive and time consuming?

A3: Not necessarily. Documentation may consist of an e-mail from such a reference or alternate source or a note to a file about an interview or phone conversation.

Q4: May I consider credit history as another source of "independent information" in evaluation of an applicant's background?

A4: Yes. However, the licensee will need to comply with applicable State laws or local ordinances that prohibit employment discrimination based on credit history. Nothing in the NRC regulations prohibits a licensee from conducting additional types of checks, and the licensee may always use measures beyond the regulatory minimum required by the access authorization program.

§ 37.25, "Background Investigations" (continued)

§ 37.25(b), "Grandfathering"

§ 37.25(b)(1)

Individuals who have been determined to be trustworthy and reliable for unescorted access to category 1 or category 2 quantities of radioactive material under the Fingerprint Orders may continue to have unescorted access to category 1 and category 2 quantities of radioactive material without further investigation. These individuals shall be subject to the reinvestigation requirement.

§ 37.25(b)(2)

Individuals who have been determined to be trustworthy and reliable under the provisions of 10 CFR Part 73 of this chapter or the security orders for access to SGI, SGI-

M handling, or risk-significant material may have unescorted access to category 1 and category 2 quantities of radioactive material without further investigation. The licensee shall document that the individual was determined to be trustworthy and reliable under the provisions of 10 CFR Part 73 of this chapter or a security order. Security order in this context refers to any order that was issued by the NRC that required fingerprints and an FBI criminal history records check for access to SGI, SGI-M handling, or risk-significant material, such as special nuclear material or large quantities of uranium hexafluoride. These individuals shall be subject to the reinvestigation requirement.

Explanation

These provisions relieve individuals who have been determined trustworthy and reliable under comparable NRC-required background investigations from further investigation.

Q&As

Q1: If I have approved an individual for unescorted access under a previously issued NRC security order, do I need to conduct a new background investigation?

A1: No. In addition, the licensee does not need to conduct a new background investigation for employees who were granted unescorted access to category 1 or category 2 quantities of radioactive material or access to SGI under legally binding requirements issued by an Agreement State or the NRC. These previously approved individuals are considered to be grandfathered and, therefore, do not need a new background investigation to meet the new requirements. However, the individuals will need to undergo a reinvestigation 10 years after the initial determination, and that background investigation will need to meet 10 CFR Part 37 requirements for the reinvestigation.

Q2: Do you have grandfather provisions for those who are long-term employees regarding trustworthiness and reliability?

A2: No. Long-term employees are not automatically grandfathered unless they have previously undergone a background investigation as discussed in A1 above or unless they fall under one of the categories of individuals granted relief from elements of the background investigation. See the Q&As on 10 CFR 37.29 for additional information on who may be relieved from fingerprinting.

Q3: I have an access authorization covered by 10 CFR Part 73. Do I need to reinvestigate individuals before I allow them to have unescorted access to category 2 or greater quantities of my material?

A3: No. If the licensee has a 10 CFR Part 73 access authorization program, it may grandfather an individual under 10 CFR 37.25(b)(2) if it documents that he or she was determined to be trustworthy and reliable under 10 CFR Part 73 or an NRC security order that requires fingerprints and an FBI criminal history records check for access to SGI, SGI-M, or risk-significant material, such as special nuclear material or large quantities of uranium hexafluoride.

§ 37.25, "Background Investigations" (continued)

§ 37.25(c), "Reinvestigations"

Licensees shall conduct a reinvestigation every 10 years for any individual with unescorted access to category 1 or category 2 quantities of radioactive material. The reinvestigation shall consist of fingerprinting and an FBI identification and criminal history records check in accordance with § 37.27. The reinvestigations must be completed within 10 years of the date on which these elements were last completed.

Explanation

This provision requires background reinvestigations every 10 years.

Q&As

Q1: Why is a reinvestigation required every 10 years?

A1: A reinvestigation every 10 years is necessary because an individual's criminal history may change over time in a manner that can adversely affect his or her T&R. The 10-year clock begins on the day an individual was given unescorted access to the radioactive material.

Q2: What elements are included in the reinvestigation?

A2: The reinvestigation is not a complete check. It is limited to fingerprinting and the FBI criminal history records check; it does not need to include identification through employment verification or the character and reputation determination. However, if the reinvestigation finds new information about a potentially adverse change in the individual's criminal records, the licensee may need to gather information from additional sources to consider whether to revise its previous determination about the individual's character and reputation.

Q3: Must I do a reinvestigation of an individual who qualified under 10 CFR 37.29 for relief from fingerprinting and background investigations, such as an individual approved under another Federal security review program?

A3: The relief provided by 10 CFR 37.29 does apply to the reinvestigation; however, the licensee will need to check and document that the individual still meets the relief category. If the affected individual has an approval that is still valid under another Federal security review program, the licensee does not need to conduct a reinvestigation. However, the regulation in 10 CFR 37.29(a)(12) requires the individual who was approved under the other Federal security review program to make available the appropriate documentation that he or she has an active Federal security clearance, and the licensee must receive written confirmation from the agency or employer that granted the Federal security clearance or reviewed the criminal history records check.

§ 37.27, "Requirements for Criminal History Records Checks of Individuals Granted Unescorted Access to Category 1 or Category 2 Quantities of Radioactive Material"

§ 37.27(a), "General Performance Objective and Requirements"

§ 37.27(a)(1)

Except for those individuals listed in 10 CFR 37.29 and those individuals grandfathered under 10 CFR 37.25(b), each licensee subject to the provisions of this subpart shall fingerprint each individual who is to be permitted unescorted access to category 1 or category 2 quantities of radioactive material. Licensees shall transmit all collected fingerprints to the Commission for transmission to the FBI. The licensee shall use the information received from the FBI as part of the required background investigation to determine whether to grant or deny further unescorted access to category 1 or category 2 quantities of radioactive materials for that individual.

Explanation

These provisions require licensees to fingerprint individuals who are to be permitted unescorted access and to submit the fingerprints to the NRC for transmission to the FBI.

Q&As

Q1: May I work directly with the FBI without having to process the fingerprints through the NRC?

A1: No. The NRC does not have the authority to allow licensees to submit fingerprints directly to the FBI instead of submitting them through the NRC. Section 149 of the AEA requires that "fingerprints obtained by an individual or entity as required [in this section] be submitted to the Attorney General of the United States through the Commission for identification and a criminal history records check." The NRC recognizes that some licensees may work directly with the FBI to process fingerprints to meet other requirements for criminal history checks. However, to meet the requirements in 10 CFR Part 37, the licensee must still process fingerprints through the NRC to the FBI unless the individual is covered under a relief category in 10 CFR 37.29.

§ 37.27, "Requirements for Criminal History Records Checks of Individuals Granted Unescorted Access to Category 1 or Category 2 Quantities of Radioactive Material" (continued)

§ 37.27(a), "General Performance Objective and Requirements" (continued)

§ 37.27(a)(2)

The licensee shall notify each affected individual that his or her fingerprints will be used to secure a review of his or her criminal history record and shall inform him or her of the procedures for revising the record or adding explanations to the record.

§ 37.27(a)(3)

Fingerprinting is not required if a licensee is reinstating an individual's unescorted access authorization to category 1 or category 2 quantities of radioactive materials if:

§ 37.27(a)(3)(i)

The individual returns to the same facility that granted unescorted access authorization within 365 days of the termination of his or her unescorted access authorization; and

§ 37.27(a)(3)(ii)

The previous access was terminated under favorable conditions.

§ 37.27(a)(4)

Fingerprints do not need to be taken if an individual who is an employee of a licensee, contractor, manufacturer, or supplier has been granted unescorted access to category 1 or category 2 quantities of radioactive material, access to SGI, or SGI-M handling by another licensee, based upon a background investigation conducted under this subpart, the Fingerprint Orders, or 10 CFR Part 73 of this chapter. An existing criminal history records check file may be transferred to the licensee asked to grant unescorted access in accordance with the provisions of 10 CFR 37.31(c).

§ 37.27(a)(5)

Licensees shall use the information obtained as part of a criminal history records check solely for the purpose of determining an individual's suitability for unescorted access authorization to category 1 or category 2 quantities of radioactive materials, access to SGI, or SGI-M handling.

Explanation

These provisions establish requirements for reinstatement of unescorted access authorizations and for the use of information from previous background investigations.

Q&As

Q1: Am I obligated to tell an individual why he or she is being fingerprinted?

A1: Yes. The regulation in 10 CFR 37.27(a)(2) requires the licensee to inform the individual that his or her fingerprints will be used to conduct a criminal history records check.

Q2: How should I inform the individual?

A2: The licensee may inform the individual orally or in writing. However, the licensee must obtain signed consent to conduct the background investigation. See also the Q&As on 10 CFR 37.23(c).

Q3: If an employee who has unescorted access quits and then returns, does the licensee need to fingerprint the individual again?

A3: If the employee has been gone for 365 days or less, had left under favorable conditions, and had previously undergone fingerprinting and an FBI criminal history records check, he or she will not need to be fingerprinted again. However, if the individual's unescorted access had been terminated for cause, the individual will need to be fingerprinted again.

Q4: Do I need to obtain fingerprints of service provider licensees and other individuals who have been granted unescorted access by the service provider or other licensee?

A4: No. The licensee does not need to obtain fingerprints of individuals employed by a service provider licensee if that licensee meets the requirements in 10 CFR Part 37. Employees of these service provider licensees who work at a customer's facility need not go through the customer's process for determining T&R for granting unescorted access. Because the service provider licensee has already made its own determination of T&R for its employees, the service provider licensee may instead provide its customers with certification of an individual's T&R for being granted unescorted access. However, the service provider licensee's program must meet the requirements of Subpart B of 10 CFR Part 37. The licensee will still need to determine that the individual should be granted unescorted access and to document its basis. The licensee is not required to accept the service provider's certification and can choose to escort the individual(s).

Q5: May I use information obtained as part of a criminal history records check for other purposes?

A5: No. The licensee may use the information obtained under this part only to determine suitability for either access to SGI or unescorted access to a category 1 or category 2 quantity of radioactive material.

§ 37.27, "Requirements for Criminal History Records Checks of Individuals Granted Unescorted Access to Category 1 or Category 2 Quantities of Radioactive Material" (continued)

§ 37.27(b), "Prohibitions"

§ 37.27(b)(1)

Licensees may not base a final determination to deny an individual unescorted access authorization to category 1 or category 2 quantities of radioactive material solely on the basis of information received from the FBI involving:

§ 37.27(b)(1)(i)

An arrest more than 1 year old for which there is no information of the disposition of the case; or

§ 37.27(b)(1)(ii)

An arrest that resulted in dismissal of the charge or an acquittal.

§ 37.27(b)(2)

Licensees may not use information received from a criminal history records check obtained under this subpart in a manner that would infringe upon the rights of any individual under the First Amendment to the Constitution of the United States, nor shall licensees use the information in any way that would discriminate among individuals on the basis of race, religion, national origin, gender, or age.

Explanation

These provisions prohibit the use of certain background investigation information.

Q&As

Q1: Why can't I deny an individual unescorted access authorization based solely on FBI information about an arrest that resulted in a dismissal or acquittal, or an arrest more than 1 year old for which there is no information on the disposition of the case?

A1: This misuse of such information would be a violation of NRC regulations and Federal law. Section 149(c)(2)(C) of the AEA requires that Commission regulations for the use of FBI criminal history records check information "ensure that no final determination may be made solely on the basis of information provided under this section involving...an arrest more than 1 year old for which there is no information of the disposition of the case; or...an arrest that resulted in dismissal of the charge or an acquittal." Taking such action on an incomplete record might unfairly penalize the individual. The charges may have possibly been dropped, or the individual may have been found innocent, and the record was never updated to reflect the additional information.

§ 37.27, "Requirements for Criminal History Records Checks of Individuals Granted Unescorted Access to Category 1 or Category 2 Quantities of Radioactive Material" (continued)

§ 37.27(c), "Procedures for Processing of Fingerprint Checks"

§ 37.27(c)(1)

For the purpose of complying with this subpart, licensees shall use an appropriate method listed in § 37.7 to submit to the U.S. Nuclear Regulatory Commission, Director, Division of Facilities and Security, 11545 Rockville Pike, ATTN: Criminal History Program/Mail Stop TWB-05 B32M, Rockville, MD 20852, one completed, legible standard fingerprint card (Form FD-258, ORIMDNRCOOOZ), electronic fingerprint scan or, where practicable, other fingerprint record for each individual requiring unescorted access to category 1 or category 2 quantities of radioactive material. Copies of these forms may be obtained by writing the Office of Information Services, U.S. Nuclear Regulatory Commission, Washington, DC 20555-0001, by calling 630-829-9565, or by e-mail to FORMS.Resource@nrc.gov.

Guidance on submitting electronic fingerprints can be found at http://www.nrc.gov/sitehelp/e-submittals.html, by calling 630-829-9565, or by e-mail to FORMS.Resource@nrc.gov. Guidance on submitting electronic fingerprints can be found at http://www.nrc.gov/site-help/e-submittals.html.

§ 37.27(c)(2)

Fees for the processing of fingerprint checks are due upon application. Licensees shall submit payment with the application for the processing of fingerprints through corporate check, certified check, cashier's check, money order, or electronic payment, made payable to "U.S. NRC." (For guidance on making electronic payments, contact the Security Branch, Division of Facilities and Security at 301-492-3531.) Combined payment for multiple applications is acceptable. The Commission publishes the amount of the fingerprint check application fee on the NRC public Web site. (To find the current fee amount, go to the Electronic Submittals page at http://www.nrc.gov/site-help/e-submittals.html and see the link for the Criminal History Program under Electronic Submission Systems.)

§ 37.27(c)(3)

The Commission will forward to the submitting licensee all data received from the FBI as a result of the licensee's application(s) for criminal history records checks.

Explanation

These provisions establish procedures for processing fingerprint checks.

Q&As

Q1: Where do I submit the fingerprints for processing?

A1: Under the AEA, the licensee is required to submit the fingerprints to the NRC, which forwards the fingerprints to the FBI for processing. If an individual comes under one of the relief categories specified in 10 CFR 37.29, the licensee will not need to submit the individual's fingerprints to the NRC. A completed fingerprint card (Form FD-258) should be sent to the U.S. Nuclear Regulatory Commission, Director, Division of Facilities and Security, 11545 Rockville Pike, ATTN: Criminal History Program/Mail Stop TWB-05B32M, Rockville, MD 20852.

Q2: How can I obtain more fingerprint cards?

A2: The licensee can request more fingerprint cards (Form FD-258) by writing to the Office of Information Services, U.S. Nuclear Regulatory Commission, Washington, DC 20555; by calling 630-829-9565; or by submitting e-mail to forms.resource@nrc.gov.

Q3: What information do I need to include on the fingerprint card?

A3: Licensees need to include the following information on each fingerprint card:

- last name, first name, and middle name
- signature of person being fingerprinted
- residence of person being fingerprinted (e.g., nearest town, State or territory, and zip code)
- date
- signature of the official taking the fingerprints
- address of the employer taking fingerprints
- reason for being fingerprinted (e.g., 10 CFR Part 37 regulatory requirement)
- aliases
- citizenship
- social security number and any of the other corresponding numbers requested on the card, if applicable (social security number only, no passport or other identification numbers) (If the individual does not have a social security number, leave this box blank.)
- date of birth
- place of birth
- sex
- race (e.g., A—Asian or Pacific Islander, B—Black, H—Hispanic/Latino, I—American Indian or Alaskan Native, or W—White)
- height
- weight
- eye color (BLK—Black, BLU—Blue, BRO—Brown, GRY—Gray, GRN—Green, HAZ—Hazel, MAR—Maroon, MUL—Multicolored, or PNK—Pink)
- hair color (BAL—Bald, BLK—Black, BLN—Blond, BLU—Blue, BRO—Brown, GRY—Gray or Partially, GRN—Green, ONG—Orange, PNK—Pink, PLE— Purple, RED—Red or Auburn, SDY—Sandy, XXX—Unknown, or WHI—White)

NRC licensees should use their NRC docket number in the field "YOUR NO. OCA." Agreement State licensees should use their two letter State abbreviation followed by a dash and the licensee's license number. For new license applications, an Agreement State may not be able to provide a license number before license issuance. In this situation, the Agreement State should create a unique identification number for the applicant to complete the fingerprinting card (e.g., CA-123456). The number must be unique for each applicant and must not be repeated once it is used. Incomplete fingerprint cards will not be processed and will be returned to the licensee.

Q4: May I use fingerprint cards I obtain from my LLEA?

A4: No. Licensees cannot use cards from other sources because of past problems with some of the cards.

Q5: Who may fingerprint my employees?

A5: The licensee should have its employees' fingerprints taken by an authorized official, such as a representative from an LLEA. An official authorized for this purpose could also be available through private entities, contractors, or an established onsite fingerprinting program. However, note that under 10 CFR 37.23(b)(2), the fingerprints of the nominated RO must be taken by a law enforcement agency, a Federal or State agency that provides fingerprinting services to the public, or a commercial fingerprinting service that has been authorized by a State to take fingerprints.

With the exception of the RO, no limitation exists on who may take the fingerprints. However, if the licensee has fingerprints taken at a facility other than that of a recognized Federal, State, or LLEA, it should ensure that the prints are taken legibly and that they match the identity of the individual named on the fingerprint card. In these cases, the individual who takes the fingerprints should, at a minimum, do the following:

- Attend training on how to take fingerprints. (The FBI offers training to take fingerprints, or such training may be available from LLEAs and some professional associations.)
- Verify the identity of the individual being fingerprinted by checking a Government-issued picture identification (e.g., a passport or driver's license), and verify that the name on the card matches the Government-issued identification.
- Sign the block on the fingerprint card labeled "SIGNATURE OF OFFICIAL TAKING THE FINGERPRINTS."

Q6: Is there a fee associated with the NRC's processing of the fingerprints?

A6: Since December 1, 2009, the fee to process each fingerprint card has been $26. Because the fee changes occasionally, the licensee should check before submitting its fee payment. The NRC publishes the amount of the fingerprint check application fee on the its public Web site. (To find the current fee amount, go to the Electronic Submittals page at http://www.nrc.gov/site-help/e-submittals.html and select the link for the Criminal History Program.) Additional fees may be charged by the entity taking the fingerprints.

Fees for the processing of fingerprint checks are due upon application. The regulation in 10 CFR 37.27(c)(2) requires licensees to submit payment with the application for the processing of fingerprints through a corporate check, certified check, cashier's check, money order, or electronic payment made payable to "U.S. NRC." (For guidance on how to make electronic payments, contact the Facilities Security Branch, Division of Facilities and Security, at 301-492-3531). Combined payment for multiple applications is acceptable.

Q7: May the NRC waive the fee for processing the fingerprints?

A7: No. The NRC by law may not waive or reduce the fee for processing the fingerprints.

Subparagraph a.(3) of Section 149 of the AEA explicitly requires the costs of an identification or records check to be paid by the individual or entity required to conduct the fingerprinting.

Q8: What method of payment does the NRC accept?

A8: The NRC's preferred method of payment is electronic payment through Pay.gov at http://www.pay.gov. The licensee can make payments through Pay.gov directly from its credit or debit card. Licensees will need to establish a password and user identification number before they can access Pay.gov. To establish an account, the licensee must send its request to paygo@nrc.gov. The request must include the licensee's name, address, point of contact, e-mail address, and phone number. The NRC will forward each request to Pay.gov, and someone from Pay.gov will contact the licensee with all of the necessary account information.

Licensees that use Pay.gov must make payments for processing before submitting applications for fingerprint checks to the NRC. Combined payment for multiple applications is acceptable. Licensees must include the Pay.gov payment receipt(s) along with the application(s).

For additional guidance on making electronic payments, contact the Facilities Security Branch, Division of Facilities and Security, at 301-492-3531. The NRC also accepts checks, cashier checks, or money orders made out to the "U.S. Nuclear Regulatory Commission" along with the submission of fingerprint cards. Fingerprint cards along with a Pay.gov receipt, check, cashier check, or money order should be sent to the Division of Facilities and Security, Mail Stop TWB-05 B32M, ATTN: Criminal History Program, U.S. Nuclear Regulatory Commission, 11545 Rockville Pike, Rockville, MD 20852-2738.

Q9: When making a payment to the NRC through Pay.gov for the processing of fingerprints, Pay.gov requires a TCN. What is a TCN, and what information should go in this field?

A9: TCN stands for "transaction control number." It is a tool that licensees can use to track their submissions, and it may include as much identifying information as would be useful for that purpose. For instance, licensees may include the names of one or more individuals for whom payment is being made, the licensee's name, and the date of submittal. Because that field is on the Pay.gov form, the licensee may use the field for individuals' names or any other identifying information.

Q10: May I submit my fingerprint cards electronically to the NRC?

A10: Yes. Some licensees may choose to make arrangements to submit fingerprints electronically to the NRC. For many licensees, however, the cost of electronic fingerprinting equipment may be prohibitive. To establish an electronic fingerprinting program with the NRC, please contact the NRC's Facility Security Branch at 301-492-3531. Please note that under 10 CFR 37.7, electronic submissions must be made in a manner that enables the NRC to receive, read, authenticate the sender, and forward the submission to the U.S. Attorney General for FBI processing. The FBI will need to certify any fingerprint equipment that will be used for the purposes of complying with the fingerprinting requirements for quality and performance standards. The NRC and FBI systems are capable of receiving electronic fingerprint transmission specification fingerprints, either Type 4 (rolled prints) or Type 14 (flat prints). A sample listing of FBI-certified fingerprint equipment is available at https://www.fbibiospecs.org/iafis/.

Q11: What happens to the fingerprint cards after the NRC receives them from the licensee?

A11: The NRC scans the fingerprint cards to transmit to the FBI electronically. The agency retains and secures the cards for approximately 1 month and then the hard-copy fingerprint information is destroyed after a month in accordance with federal guidelines. The NRC system keeps a record of all the submissions, but can only produce a copy of that record for a year; after a year, NRC cannot print out a copy.

Q12: Why might a fingerprint card be unclassifiable?

A12: Fingerprints may be unclassifiable for a number of reasons, including the following:

- Incomplete impressions were taken. (For example, the fingers were not completely rolled from one side of the nail to the other.)
- The left and right hands were reversed on the fingerprint card.
- The same hand or finger was printed twice on the card.

- The fingerprints are not clear and distinct (e.g., smudged, uneven, or too dark or light).
- The fingers on the card are missing or are partially missing without an explanation.

To avoid rejection of fingerprints by the FBI as "unclassifiable," the individual who takes the prints should ensure that they are of good quality and that they do not include any of these deficiencies and should also follow the instructions on the back of the fingerprint card. Fingerprint cards with incomplete or missing information will be returned to the licensee to provide complete information, thus resulting in a delay in processing.

The FBI has provided guidance on taking fingerprints for submission to the agency at http://www.fbi.gov/hq/cjisd/takingfps.html. This guidance also discusses special situations, such as fingerprinting an individual with abnormalities of the fingers, thumbs, or hands, and the appropriate way to identify such situations on the fingerprint card. The guidance also includes a checklist to verify that the fingerprint impressions meet the FBI's requirements.

Q13: What are the next steps in the process if the FBI rejects a Form FD-258 (fingerprint card) because the fingerprints are not classifiable? What options are available to licensees if, after multiple attempts, an individual's fingerprints cannot be classified based on conditions other than poor quality?

A13: If the initial fingerprint submission is returned by the FBI because the fingerprint impressions cannot be classified, the fingerprints may be retaken and resubmitted (i.e., a new Form FD-258 via electronic submission) for a second attempt. The licensee will not be charged for the resubmission if the licensee provides a copy of the FBI TCN or the FBI response indicating that the fingerprints could not be classified. If the FBI is unable to classify the second submission of fingerprints, the licensee may submit additional fingerprint impressions for the individual as follows:

1. The third fingerprint card submission will require payment of an additional $26 processing fee.
2. If the third submission is also returned as unclassifiable, the licensee may submit a fourth set of fingerprints. An additional fee is not required because the fee for the third submission includes one resubmission. As with the second submission, the FBI response or TCN should be included, or the submission may be treated as a new request and an additional fee may be charged. Please note that a licensee may opt to take and submit the third and fourth sets of fingerprints together to avoid a potential delay in the response. If the third set is returned as unclassifiable, the NRC will automatically resubmit the fourth set.
3. No further submissions will be required, and the licensee may consider the Results of the name search-FBI identification and criminal history records check as a component in determining trustworthiness and reliability.

The NRC will consider licensee requests for deviation from the above process for good cause (e.g., a demonstrated history of difficulty providing classifiable fingerprints during other fingerprinting programs or a documented medical condition or physical anomaly that can prevent the taking of readable prints). Licensees may submit a request for consideration of alternatives and provide the basis for the need for an alternative process to the NRC's

Facilities Security Branch in the Division of Facilities and Security (requests may be made by phone at 301-492-3531 or mailed to:

Office of Administration
Division of Facilities and Security
Mail Stop TWB-05 B32M
U.S. Nuclear Regulatory Commission
Washington, DC 20555-0012

Please note that requests for an alternative to the above process will not affect a licensee's responsibility to fingerprint individuals for unescorted access or to comply with the trustworthiness and reliability requirements.

§ 37.29, "Relief from Fingerprinting, Identification, and Criminal History Records Checks and Other Elements of Background Investigations for Designated Categories of Individuals Permitted Unescorted Access to Certain Radioactive Materials or Other Property"

§ 37.29(a)

Fingerprinting, and the identification and criminal history records checks required by Section 149 of the AEA, as amended, and other elements of the background investigation are not required for the following individuals prior to granting unescorted access to category 1 or category 2 quantities of radioactive materials:

§ 37.29(a)(1)

An employee of the Commission or of the Executive Branch of the U.S. Government who has undergone fingerprinting for a prior U.S. Government criminal history records check;

§ 37.29(a)(2)

A member of Congress;

§37.29(a)(3)

An employee of a member of Congress or Congressional committee who has undergone fingerprinting for a prior U.S. Government criminal history records check;

§ 37.29(a)(4)

The Governor of a State or his or her designated State employee representative;

§ 37.29(a)(5)

Federal, State, or local law enforcement personnel;

§ 37.29(a)(6)

State Radiation Control Program Directors and State Homeland Security Advisors or their designated State employee representatives;

§ 37.29(a)(7)

Agreement State employees conducting security inspections on behalf of the NRC under an agreement executed under Section 274.i. of the Atomic Energy Act;

§ 37.29(a)(8)

Representatives of the International Atomic Energy Agency (IAEA) engaged in activities associated with the U.S./IAEA Safeguards Agreement who have been certified by the NRC;

§ 37.29(a)(9)

Emergency response personnel who are responding to an emergency;

§ 37.29(a)(10)

Commercial vehicle drivers for road shipments of category 2 quantities of radioactive material;

§ 37.29(a)(11)

Package handlers at transportation facilities, such as freight terminals and railroad yards;

§ 37.29(a)(12)

Any individual who has an active Federal security clearance, provided that he or she makes available the appropriate documentation. Written confirmation from the agency/employer that granted the Federal security clearance or reviewed the criminal history records check must be provided to the licensee. The licensee shall retain this documentation for a period of 3 years from the date the individual no longer requires unescorted access to category 1 or category 2 quantities of radioactive material; and

§ 37.29(a)(13)

Any individual employed by a service provider licensee for which the service provider licensee has conducted the background investigation for the individual and approved the individual for unescorted access to category 1 or category 2 quantities of radioactive material. Written verification from the service provider must be provided to the licensee. The licensee shall retain the documentation for a period of 3 years from the date the individual no longer requires unescorted access to category 1 or category 2 quantities of radioactive material.

Explanation

This section identifies the categories of individuals who may be relieved from the background investigation elements. In addition, it establishes the conditions under which a licensee may accept a Federal agency security clearance or the T&R determination of a service provider licensee.

Q&As

Q1: What is the basis for relieving individuals from fingerprinting and criminal history records check requirements?

A1: Under Section 149.b of Chapter 12 of the AEA, the NRC may, by rule, relieve individuals from the fingerprinting, identification, and criminal history records check requirements if it finds that such action is "consistent with its obligations to promote the common defense and security and to protect the health and safety of the public."

Q2: How may I determine if an individual has undergone a previous background investigation and falls under one of the categories listed in 10 CFR 37.29?

A2: For some categories of individuals, it will be obvious. For example, a police officer, an NRC employee, or State employee will have identification of some sort that identifies the individual and his or her employer so that no further checking will be necessary. For those categories of individuals for which it is not obvious, either the individual or the individual's employer will need to provide documentation.

Q3: What type of documentation is necessary?

A3: The regulation in 10 CFR 37.29(a)(12)) requires documentation to consist of written confirmation from the agency or employer that granted a Federal security clearance or reviewed the criminal history records check. Documentation may be a confirmation letter that the named individual has a security clearance that meets the requirements of Subpart B and that was issued on a specified date. Documentation may also consist of a copy of the background investigation determination itself if the agency or employer provides it. To show that the service provider's employee has undergone a background check under 10 CFR 37.29(a)(13), which is equivalent to that required under 10 CFR 37.25, the service provider licensee must provide the service recipient licensee a written communication that includes the name and identifying information of the employee who will be providing the service and an affirmation that this employee has been determined to be trustworthy and reliable in accordance with 10 CFR 37.25(a)(1) through 10 CFR 37.25(a)(7). To comply with these requirements, the service provider's background investigation must include fingerprinting and an FBI identification and criminal history records check; verifications of the individual's true identity, employment history, and education; and a character and reputation determination.

Q4: How long must I keep documentation records?

A4: The regulation in 10 CFR 37.23(h)(1) requires the licensee to keep the records for at least 3 years after the individual no longer requires unescorted access to category 1 or category 2 quantities of radioactive material or, under 10 CFR 37.103, "Record Retention," until its license is terminated, whichever comes first.

Q5: May emergency first responders, such as police and fire department personnel, be deemed trustworthy and reliable without a background check?

A5: Yes. The NRC, State radiation protection agencies, and local law enforcement officials are relieved from the background investigation elements for purposes of this requirement. In the event of an emergency, such as a fire or explosion, firefighters may be granted unescorted access for the purposes of controlling the emergency situation. Firemen are not provided relief from the background investigation for routine inspections that the fire department may conduct.

Q6: May properly qualified security equipment vendors and/or service providers be considered trustworthy and reliable and granted unescorted access to the radioactive material or devices containing the radioactive material?

A6: Yes. However, the vendor/service provider must meet the requirements in 10 CFR Part 37. Employees of the vendor/service provider licensees do not need to go through the customer's process for determining their T&R for unescorted access at a customer's facility; instead, if the vendor/service provider licensee has made its own determination of T&R for its personnel, this licensee may provide its customers a certification of each employee's T&R for being granted unescorted access. However, the vendor or service provider's program must meet the requirements in Subpart B of 10 CFR Part 37.

Q7: What does the NRC mean by "commercial driver" in 10 CFR 37.29(a)(10)?

A7: A commercial driver is someone who drives commercial vehicles for a living. For example, someone driving for FedEx or the United Parcel Service (UPS) is considered a commercial driver. The NRC will rely on DOT and Transportation Security Administration (TSA) programs for background investigations of these personnel. Note that this relief is only provided for individuals transporting category 2 quantities of radioactive material. A commercial driver transporting a category 1 quantity will need to undergo a background investigation. The NRC will not consider an individual who works for the licensee and drives a company truck between radiography jobs to be a commercial driver, and he or she will need to undergo a background investigation before having unescorted access to a category 1 or category 2 quantity of radioactive material.

Q8: Why are commercial drivers and package handlers relieved from undergoing background checks under 10 CFR Part 37?

A8: These individuals are typically outside the control of the licensee, and it would have no way of knowing or influencing who those individuals might be. The NRC will rely on DOT and TSA programs for background investigations of these individuals. Although the background investigation may not be identical to those required under 10 CFR Part 37, the NRC believes that the potential risk that a commercial driver or package handler might pose due to any difference in the background investigation is acceptably small.

Q9: If an individual falls under one of the categories listed for relief in 10 CFR 37.29, am I required to grant the individual unescorted access to the material?

A9: No. The regulation in 10 CFR Part 37 does not require the licensee to grant unescorted access to any radioactive materials or other property subject to NRC regulations to any individual. The licensee still needs to decide whether to grant or deny an individual unescorted access independent of if he or she qualifies for relief from fingerprinting and other background investigation elements under one of the categories in 10 CFR 37.29. This section simply clarifies that a licensee may permit unescorted access to certain categories of individuals without performing a background investigation if these officials are otherwise qualified for access and have a demonstrated need for it. However, the licensee will need to conduct any training necessary under 10 CFR 37.43(c) before granting such individuals unescorted access.

Q10: Would State radiation control program directors or their designated State employee representatives be exempt from background checks under 10 CFR Part 37 even if their need for unescorted access is not security related, such as in a safety inspection?

A10: Yes. The regulation in 10 CFR 37.29(a)(6) relieves these officials from the background investigation elements.

Q11: Am I required to allow an NRC or Agreement State inspector unescorted access?

A11: No. The regulations in 10 CFR 37.105, "Inspections," which provide inspectors access to a licensee's facilities and records, do not require the licensee to provide unescorted access. The licensee may always escort an inspector when he or she is in a security zone or is inspecting protected information.

§ 37.29, "Relief from Fingerprinting, Identification, and Criminal History Records Checks and Other Elements of Background Investigations for Designated Categories of Individuals Permitted Unescorted Access to Certain Radioactive Materials or Other Property" (continued)

§ 37.29(b)

Fingerprinting, and the identification and criminal history records checks required by Section 149 of the AEA, as amended, are not required for an individual who has had a favorably adjudicated U.S. Government criminal history records check within the last 5 years, under a comparable U.S. Government program involving fingerprinting and an FBI identification and criminal history records check provided that he or she makes available the appropriate documentation. Written confirmation from the agency/employer that reviewed the criminal history records check must be provided to the licensee. The licensee shall retain this documentation for a period of 3 years from the date the individual no longer requires unescorted access to category 1 or category 2 quantities of radioactive material. These programs include, but are not limited to:

§ 37.29(b)(1)

National Agency Check;

§ 37.29(b)(2)

Transportation Worker Identification Credential (TWIC) under 49 CFR 1572;

§ 37.29(b)(3)

Bureau of Alcohol, Tobacco, Firearms, and Explosives background check and clearances under 27 CFR 555;

§ 37.29(b)(4)

Health and Human Services security risk assessments for possession and use of select agents and toxins under 42 CFR 73;

§ 37.29(b)(5)

Hazardous material security threat assessment for hazardous material endorsement to commercial drivers license under 49 CFR 1572; and

§ 37.29(b)(6)

Customs and Border Protection's Free and Secure Trade (FAST) Program.

Explanation

Individuals with a favorably adjudicated U.S. Government criminal history records check under a comparable Federal program within the last 5 years may be relieved from the 10 CFR Part 37 fingerprinting and FBI criminal history records check. However, the licensee must conduct the other elements of the required background investigation.

Q&As

Q1: What does it mean for an individual to have a "favorably adjudicated" U.S. Government criminal history records check?

A1: The licensee may consider an individual's criminal history records check to have been "favorably adjudicated" if the individual is authorized under the applicable program to conduct the activity for which the check was made. The regulation in § 37.29(b) states that the adjudication must have been conducted within the last 5 years under a comparable U.S. Government program involving fingerprinting and an FBI identification and criminal history records check, such as those programs listed.

Q2: What does [the] NRC consider a comparable U.S. Government program?

A2: A comparable U.S. Government program is one that requires an individual to submit fingerprints for an FBI criminal history records check. In most cases, the Government agency would issue an approval that could include an identification card or a specific approval to work in certain areas. Comparable U.S. Government programs include the following:

- National Agency Check
- Transportation Worker Identification Credential
- Bureau of Alcohol, Tobacco, Firearms and Explosives background check and clearances under 27 CFR 555, "Commerce in Explosives"
- U.S. Department of Health and Human Services security risk assessments for possession and use of select agents and toxins under 42 CFR 73, "Select Agents and Toxins"
- hazardous material security threat assessment of a hazardous material endorsement to a commercial driver's license under 49 CFR 1572, "Credentialing and Security Threat Assessments"
- U.S. Customs and Border Protection's Free and Secure Trade Program

Q3: What documentation must I have in order to accept the individual's claim that he or she has been favorably adjudicated?

A3: The regulation in 10 CFR 37.29(b) requires the individual to make available documentation that his or her adjudication is currently valid, and the licensee must receive written confirmation from the agency or employer that reviewed the criminal history records check.

Q4: May I grant unescorted access to category 2 or greater quantities of radioactive materials to an individual relieved from fingerprinting and criminal history investigation without any other checks?

A4: No. The licensee must still fulfill the requirements in 10 CFR 37.25(a)(2) through 10 CFR 37.25(a)(6) for other elements of the background investigation. These elements include a verification of true identity; a verification of employment and education; a character and reputation determination; and an acquisition and, to the extent possible, an evaluation of information obtained from sources independent of those provided by the individual. After completing these checks, the licensee may then grant unescorted access if the individual's job duties require it.

Q5: When an individual has been favorably adjudicated under a "comparable U.S. Government program" under 10 CFR 37.29(b), does this mean comparable to one of the programs listed under 10 CFR 37.29(b)(1) through (b)(6) above, or comparable to the background investigation requirements of 10 CFR Part 37?

A5: To be "comparable" for the purposes of 10 CFR 37.29(b), the program must include fingerprinting and an FBI criminal history records check.

Q6: Do I need to obtain a signed consent from an individual who has undergone a fingerprinting and a criminal history records check under a comparable Federal program?

A6: Yes. The licensee will need signed consent from an employee who has undergone fingerprinting and a criminal history records check under a comparable Federal program because it will still need to conduct the other elements of the background investigation required by 10 CFR 37.25. These other elements include verification of the individual's true identity, employment, and education history and an evaluation of his or her character and reputation.

Q7: Must I do a reinvestigation of an individual who qualified for relief from fingerprinting and background investigations, such as an individual approved under another Federal security review program under 10 CFR 37.29(b)?

A7: The relief provided by 10 CFR 37.29 does apply to the reinvestigation, but the licensee will need to check and document that the individual still meets the relief category. If the affected individual has an approval that is still valid under another Federal security review program, the licensee does not need to conduct a reinvestigation.

§ 37.31, "Protection of Information"

§ 37.31(a)

Each licensee who obtains background information on an individual under this subpart shall establish and maintain a system of files and written procedures for protection of the record and the personal information from unauthorized disclosure.

Explanation

Licensees must establish and maintain a filing system and written procedures to protect from unauthorized disclosure records and personal information produced from background investigations under 10 CFR Part 37.

Q&As

Q1: Are licensees required to protect information obtained during a background investigation?

A1: Yes. The collected information will likely contain personally identifiable information (PII) and should be provided only to authorized individuals. Under Section 149(c)(2)(D) of the AEA, as amended, the Commission is required to prescribe requirements for the use of background investigation information "to protect individuals subject to fingerprinting under this section from misuse of the criminal history records." Accordingly, under 10 CFR 37.31(a), the licensee is required to establish and maintain a system of files and procedures to protect the information from disclosure to any unauthorized person. The licensee should, as good practice, store background investigation documentation in a locked drawer or file cabinet.

Q2: What does the NRC consider to be "unauthorized disclosure"?

A2: The NRC considers "disclosure" to be the providing of, either deliberately or inadvertently, any information obtained in a background investigation in accordance with this subpart by any means, including electronic means, such as facsimile, voice mail, or e-mail. Such disclosure is "unauthorized" if the recipient of the information is not the subject individual, the individual's representative, an authorized representative of the NRC or an Agreement State agency, or an RO. (See 10 CFR 37.31(b).)

Q3: If I need to prepare an information protection program for my employees' background investigation information, and I want someone other than the RO or other physical security staff (e.g., an information technology specialist) to prepare the program, do I have to conduct a full background investigation on that individual, even if he or she does not require access to radioactive material to perform a job duty?

A3: No. An individual who prepares an information protection program for the access authorization program is not required to undergo a background investigation.

Q4: Should my procedures for information protection address the destruction of records?

A4: This section does not explicitly require that a licensee's information protection procedures address the destruction of records, but this procedure would be a good practice, especially for PII. PII is information that can be used, by itself or in combination with other information, to distinguish or trace an individual's identity. It includes personal information, such as an individual's social security number; date and place of birth; mother's maiden name; biometric records; and records of his or her education, financial transactions, and employment or medical history. To prevent an unauthorized individual from gaining access to records containing PII, the licensee should destroy them by shredding, burning, pulping, degaussing, or using a similarly irreversible method. For more information and guidance on protecting PII, an OMB memorandum to Federal agencies entitled, "Safeguarding Against and Responding to the Breach of Personally Identifiable Information," dated May 22, 2007, may be useful. The licensee should also check for applicable State or local government requirements for PII protection and destruction.

§ 37.31, "Protection of Information" (continued)

§ 37.31(b)

The licensee may not disclose the record or personal information collected and maintained to persons other than the subject individual, his or her representative, or to those who have a need to have access to the information in performing assigned duties in the process of granting or denying unescorted access to category 1 or category 2 quantities of radioactive material, *SGI*, or *SGI-M* handling. No individual authorized to have access to the information may disseminate the information to any other individual who does not have a need to know.

Explanation

This provision restricts the disclosure or the re-dissemination of personal information collected under 10 CFR Part 37 background investigations.

Q&As

Q1: Under what circumstances may I disclose the personal information obtained during a background investigation?

A1: The licensee may disclose the information to the subject individual, his or her representative, a duly authorized representative of the NRC or an Agreement State agency, or an individual with a need for access to the information to perform an assigned duty that supports a decision on unescorted access to SGI, SGI-M, or a quantity of radioactive material subject to 10 CFR Part 37. If the licensee has a question about whether an individual has a need to know the information, it should verify the need before providing the information. A supervisor should be contacted to ensure that the request is legitimate.

The licensee should, as good practice, ask the requestor to put the request in writing and to specify the nature of the information requested and should document the requestor's need to know. State, local, or Federal law may also require disclosure for such purposes as criminal investigations.

Q2: How do I determine that another person requesting access to these records or personal information has a need to know?

A2: If it is not the individual or an NRC or Agreement State inspector, the licensee should inquire why the individual needs the information. The licensee may contact the individual's supervisor to verify the need for the information. If the individual is from another company, the licensee must find out why he or she is requesting the information. In addition, the licensee must ensure that the request is legitimate and must determine if it will comply with the request. The licensee is not required to share information with other licensees.

Q3: Do I need to document a need to know?

A3: The regulations in 10 CFR Part 37 contain no specific requirements to document a need to know. The licensee, however, should consider this a prudent practice to enable it to demonstrate compliance with the requirement not to disseminate the information to any other individual who does not have a need to know.

> **§ 37.31, "Protection of Information" (continued)**
>
> **37.31(c)**
> The personal information obtained on an individual from a background investigation may be provided to another licensee:
>
> **37.31(c)(1)**
> Upon the individual's written request to the licensee holding the data to disseminate the information contained in his or her file; and
>
> **37.31(c)(2)**
> The recipient licensee verifies information, such as name, date of birth, social security number, gender, and other applicable physical characteristics.

Explanation

These provisions establish conditions under which a licensee may provide personal information that it obtained on an individual from a background investigation to another licensee.

Q&As

Q1: May one licensee transfer personal information it obtained during an investigation to another licensee?

A1: Yes, the licensee may do so if the subject individual makes a written request to the licensee to transfer the information contained in his or her file.

Q2: If I receive background investigation information from another licensee, may I rely on that information, or must I re-verify it?

A2: Unless the information is older than the 10-year reinvestigation period, the licensee may rely on the information without re-verification. However, the regulation in 10 CFR 37.23(e)(3) requires the licensee to document the basis for concluding that each individual with unescorted access to category 2 or greater quantities of radioactive material is trustworthy and reliable. The receiving licensee must verify information, such as the name, date of birth, social security number, gender, and other physical characteristics of the subject individual, to ensure that he or she is the person whose file has been transferred. The receiving licensee must ensure that the information transferred is for the correct individual. The receiving licensee may ask the originating licensee to provide information about distinguishing characteristics so that it can verify the individual. Such characteristics could include estimated height; birthmarks; scars; tattoos; missing or partially missing fingers or fingernails; and, if permissible under applicable laws and regulations, race and ethnicity.

Q3: Must I re-disseminate an employee's background investigation information if he or she requests it?

A3: No. The regulations in 10 CFR Part 37 contain no such requirement. Sharing with another licensee the information that a licensee developed is entirely at its discretion.

§ 37.31, "Protection of Information" (continued)

§ 37.31(d)

The licensee shall make background investigation records obtained under this subpart available for examination by an authorized representative of the NRC to determine compliance with the regulations and laws.

§ 37.31(e)

The licensee shall retain all fingerprint and criminal history records (including data indicating no record) received from the FBI, or a copy of these records if the individual's file has been transferred, on an individual for 3 years from the date the individual no longer requires unescorted access to category 1 or category 2 quantities of radioactive material.

Explanation

Licensees must make background investigation records available to the NRC for its examination to determine compliance. Licensees must also retain all fingerprint and criminal history information that it received from the FBI on an individual for 3 years from the date that the individual no longer requires unescorted access.

Q&As

Q1: Does the NRC have a right to review the background investigation records?

A1: Yes. To determine compliance with applicable laws and regulations, an inspector or other authorized representative of the NRC may examine the background investigation records, and 10 CFR 37.31(d) requires the licensee to make them available for examination.

Q2: What background investigation records do I need to maintain for NRC inspection under 10 CFR 37.31(d)?

A2: This subsection requires the licensee to make available for the NRC's examination "background investigation records obtained under this subpart." This requirement would include, among other things, all records that must be maintained under 10 CFR 37.23(e) concerning the licensee's basis for concluding if "reasonable assurance that an individual is trustworthy and reliable" exists. In addition, 10 CFR 37.23(h)(1) more generally requires the licensee to retain for 3 years "documentation regarding the T&R of individual employees." The NRC considers this documentation to include, in addition to the fingerprint and criminal history records covered under 10 CFR 37.27(e) above, all background investigation records that the licensee must obtain under 10 CFR 37.25(a)(2) through 10 CFR 37.25(a)(6). In addition, 10 CFR 37.29(b) requires the licensee to retain any written confirmations that it received concerning a favorably adjudicated criminal history records check, 10 CFR 37.29(a)(12) requires the licensee to retain any written confirmations that it received from U.S. Government entities concerning a Federal security clearance, and the regulation in 10 CFR 37.29(a)(13) requires the licensee to retain any written verifications that it received from service providers.

Q3: What fingerprint and criminal history records do I need to maintain?

A3: The regulation in 10 CFR 37.31(e) requires the licensee to retain all fingerprint and criminal history records (including data that indicate no record) that it received from the FBI or a copy of these records if the individual's file has been transferred.

Q4: How long do I need to keep background investigation and fingerprint and criminal history records on an individual?

A4: The regulation in 10 CFR 37.31(e) requires that if the licensee receives an individual's fingerprint and criminal history file from another party, it must retain a copy of these records for 3 years from the date that the individual no longer requires unescorted access to category 1 or category 2 quantities of radioactive material. The regulation in 10 CFR 37.23(h)(1) requires the licensee to retain records on its determination of an individual's T&R, including background investigation records, for 3 years from the date that the individual no longer requires unescorted access. However, if the licensee's license is terminated before either of these retention periods elapses, the licensee, under 10 CFR 37.103, only needs to retain these records until the termination of its license.

Q5: According to 10 CFR 37.23(h), I need to keep records for 3 years. Since these records will contain information that needs to be protected, how do I properly dispose of them?

A5: The background investigation records will likely contain personal information that will be considered PII. Records containing PII should be destroyed, not just tossed in the trash, to prevent an unauthorized individual from gaining access to that information.

§ 37.33, "Access Authorization Program Review"

§ 37.33(a)

Each licensee shall be responsible for the continuing effectiveness of the access authorization program. Each licensee shall ensure that access authorization programs are reviewed to confirm compliance with the requirements of this subpart and that comprehensive actions are taken to correct any noncompliance that is identified. The review program shall evaluate all program performance objectives and requirements. Each licensee shall periodically (at least annually) review the access program content and implementation.

§ 37.33(b)

The results of the reviews, along with any recommendations, must be documented. Each review report must identify conditions that are adverse to the proper performance of the access authorization program, the cause of the condition(s), and, when appropriate, recommend corrective actions and corrective actions taken.

The licensee shall review the findings and take any additional corrective actions necessary to preclude repetition of the condition, including reassessment of the deficient areas where indicated.

§ 37.33(c)
Review records must be maintained for 3 years.

Explanation

Licensees must review and assess the effectiveness of their access authorization programs, take appropriate corrective actions, and maintain the records for 3 years.

Q&As

Q1: How should I evaluate the "effectiveness" of my access authorization program to comply with NRC requirements for program reviews?

A1: The licensee should consider several things when evaluating the continued effectiveness of its program. Specifically, the licensee should consider its ability to demonstrate compliance with all the applicable requirements in this subpart and to take comprehensive and effective actions to correct identified non-compliances. Most importantly, the licensee should keep in mind that continuing effectiveness is not a static condition. Continuing improvements are an essential part of an effective program.

To minimize the potential for overlooking an adverse condition, the licensee's program reviews should address each applicable requirement of this subpart. These reviews should identify adverse conditions, non-compliances, and root causes and should provide corrective actions. The licensee also should follow up on the implementation of these actions and reassess their effect on the program. The hallmark of an effective program is not the absence of recorded adverse conditions or non-compliances; it is documented evidence that the licensee has made a diligent effort to find these problems and that it is continuing to reassess the effectiveness of its actions to prevent problems from reoccurring. Ultimately, the most important indicator of an effective program is that the licensee is consistently able to identify and successfully address existing and emerging deficiencies.

Q2: How do I ensure that I will meet the requirement to review the access authorization program "at least annually"?

A2: Recognizing that some demands on a licensee's time and resources are beyond its control, the NRC will consider that it is conducting a program review "at least annually" if it conducts such a review each year at about the same time of year, not to exceed 12 months.

Q3: Must I engage an outside contractor to conduct access authorization program reviews?

If not, how may I ensure that the review isn't conducted by the same people carrying out the activities being reviewed?

A3: Although hiring an independent party to conduct access authorization program reviews would be one way to demonstrate compliance, the regulation does not require it.

Although self assessments do have value and would meet the requirement, the licensee should try to avoid a situation in which individuals are reviewing their own work. If the licensee has a large enough staff, it could establish a review team of approved individuals to avoid self reviews. This team could be led by an individual, such as the security official or RSO, who works outside the management chain of the licensee's access authorization staff. If the licensee has separate programs at more than one location with a different RO at each location, another way to run a more arm's length review would be to have the review team at one location review the program implemented by the staff of a different facility. Licensees also may choose to set up a review team through an industry association using participants from several independent member organizations to conduct program reviews.

Q4: Do I need to report to the NRC any noncompliance identified during an access authorization program review?

A4: No. The licensee does not have to report any noncompliance to the NRC unless a specific regulation required such a report to the agency.

Q5: What would the NRC consider a "condition adverse to the proper performance of the access authorization program"?

A5: The NRC will consider an adverse condition to be anything that, if not corrected, could impair the effectiveness of the licensee's access authorization program or its continuing compliance with the requirements of this subpart. An example of an adverse condition might be a delay in the reevaluation of an employee's T&R after receiving information about a felony arrest. Another example might be a failure to determine beforehand, under 10 CFR 37.31(b), the need to know of an unfamiliar individual who makes an initial request for the re-dissemination of personal information about an employee. An adequate program review should never be limited to looking only for existing or imminent non-compliances. It should assess or reassess all conditions that may call into question the continuing effectiveness of the licensee's access authorization program.

Q6: 10 CFR 37.33(b) requires the report resulting from a program review to recommend corrective actions "when appropriate." When should I recommend corrective actions?

A6: At a minimum, a program review report should recommend one or more corrective actions for each noncompliance or condition "adverse to the proper performance of the access authorization program" identified as a result of the review.

Q7: What should I consider as "review documentation" for the purposes of this subsection?

A7: The licensee should retain the annual review report itself and any attachments or enclosures related to that report. Related records should include the membership and leadership of the review team, if applicable; a description of the management approval process for the annual report, if applicable; root cause analyses for identified non-compliances or adverse conditions; recommended corrective actions; evaluations of the effectiveness of past corrective actions; and other documents that were considered in the review. A good review documentation practice is to include minority views on issues in the report for which significant professional disagreement is present. The NRC does not expect the licensee to retain rough drafts of its annual access authorization program reviews, its meeting records, or the notes of each member of a review team.

Q8: How long should I maintain records of the program review?

A8: Records must be maintained for 3 years under 10 CFR 37.33(c).

Q9: Do I need to keep a paper copy of the program review or may I keep an electronic copy?

A9: The licensee may keep either a paper copy or an electronic copy as long as the record is legible for the required period and can be accessed. To discourage unauthorized alteration, the licensee should make electronic copies read-only if possible, such as by saving them in a PDF.

ANNEX A. ADDITIONAL GUIDANCE FOR EVALUATING AN INDIVIDUAL'S TRUSTWORTHINESS AND RELIABILITY FOR ALLOWING UNESCORTED ACCESS TO CERTAIN RADIOACTIVE MATERIAL

Each licensee is responsible for determining whether to grant an individual unescorted access to certain radioactive materials. The licensee shall allow only individuals who it has approved and documented as trustworthy and reliable to have unescorted access to a category 2 or greater quantity of radioactive material and devices containing that quantity. The licensee's reviewing official (RO) makes the trustworthiness and reliability (T&R) determination based on the information collected during the background investigation. The RO must document the determination basis. Note that the guidance also can be used when information is received after an individual has been approved for unescorted access.

Unescorted access determinations require an evaluation of a person's T&R. When a person's life history shows evidence of unreliability or untrustworthiness, a licensee may question if that person can be relied on and trusted to exercise the responsibility necessary for working with risk-significant radioactive materials. The purpose of the T&R determination requirement is to provide reasonable assurance that those individuals are trustworthy and reliable and do not constitute an unreasonable risk to the public health and safety, including the potential to commit or aid theft or radiological sabotage. In evaluating the relevance of an individual's conduct, the RO should consider the following factors:

- the nature, extent, and seriousness of the conduct
- the circumstances surrounding the conduct, including evidence as to if it was deliberate
- the frequency and recency of the conduct
- the individual's age and maturity at the time of the conduct
- the extent to which participation in the conduct was voluntary
- the presence or absence of rehabilitation and other permanent behavioral changes
- the motivation for the conduct
- the potential for pressure, coercion, exploitation, or duress as a result of the conduct
- the likelihood of continuation or recurrence

Each case must be judged on its own merits, and the final determination remains the responsibility of the licensee. In every case, the RO must evaluate T&R based on an accumulation of information that supports a positive finding before granting unescorted access. Items to consider include the following:

- if the information collected is consistent and adequate
- if the applicant's true identity can be reasonably verified by comparing applicant provided identification and personal history data to pertinent information from the background investigation and other data sources
- if inconsistencies identified by the licensee's review or investigation are intentional, innocent, or an oversight

Willful or intentional acts of omission or untruthfulness could be grounds for denial of unescorted access. When a licensee submits fingerprints to the NRC, it will receive an FBI identification and criminal history record since the individual's 18th birthday. The licensee's RO should evaluate that information using the guidance below. The licensee's RO is required to evaluate all available information in making a T&R determination for unescorted access to radioactive materials, including the criminal history records information pertaining to the individual. The FBI identification and criminal history records check is used to determine if the individual has a record of criminal activity that indicates that the individual should not have unescorted access to radioactive materials subject to the requirements in 10 CFR Part 37. The licensee's documentation of each T&R determination for unescorted access to radioactive materials, which must include a review of criminal history information, must include the basis for the decision made. Licensees should not make a final determination based solely on criminal history check information involving an arrest of more than 1 year old for which no information is available on the disposition of the case or an arrest that resulted in dismissal or acquittal of the charge.

The criminal history records check is used to evaluate if the individual has a record of criminal activity that may compromise his or her T&R. Identification of a criminal history through the FBI criminal history records check or a discretionary local criminal history check does not automatically indicate unreliability or untrustworthiness of the employee. The licensee will have to judge the nature and recency of the criminal activity. The licensee may authorize individuals with criminal records for unescorted access to radioactive materials based on a documented evaluation of the basis for determining that the employee or applicant is reliable and trustworthy, notwithstanding his or her criminal history. The licensee's documentation of each review of criminal history and other background checks information must include the decision-making basis. When evaluating the results of the criminal history records check, the licensee should consider, at a minimum, if the subject individual did the following:

- Committed, attempted to commit, aided, or abetted another individual who committed or attempted to commit any act of sabotage, espionage, treason, sedition, or terrorism.
- Publicly or privately advocated actions that may be inimical to the interest of the United States, or publicly or privately advocated the use of force or violence to overthrow the Government of the United States or the alteration of the form of government of the United States by unconstitutional means.
- Knowingly established or continued a sympathetic association with a saboteur, spy, traitor, seditionist, anarchist, terrorist, or revolutionary; with an espionage agent or other secret agent or representative of a foreign nation whose interests may be inimical to the interests of the United States; or with any person who advocates the

use of force or violence to overthrow the Government of the United States or the alteration of the form of government of the United States by unconstitutional means. (Ordinarily, the licensee should not consider chance or casual meetings or contacts limited to normal business or official relations.)
Knowingly joined or engaged in any activity in sympathy with, or in support of, any foreign or domestic organization, association, movement, group, or combination of persons who advocate or practice the commission of acts of force or violence to prevent others from exercising their rights under the Constitution or laws of the United States or any State or any subdivisions thereof by unlawful means or who advocate the use of force and violence to overthrow the Government of the United States or the alteration of the form of government of the United States by unconstitutional means. (Ordinarily, the licensee should not consider chance or casual meetings or contacts limited to normal business or official relations.)

- Deliberately misrepresented, falsified, or omitted relevant and material facts from documentation provided to the licensee.
- Had been convicted of a crime(s) that, in the RO's opinion, indicates poor judgment, unreliability, or untrustworthiness.

Licensees also may consider how recently such indicators occurred and other extenuating or mitigating factors in their determinations. Section 149.c.(2)(B) of the Atomic Energy Act of 1954, as amended, requires that the information obtained as a result of fingerprinting be used solely for the purposes of making a determination as to suitability for unescorted access. Such a determination is not a hiring decision, and the NRC does not intend for licensees to use this guidance for such purposes. A determination that a particular individual is unsuitable for unescorted access does not necessarily mean that he or she is unsuitable for escorted access or for some other position that does not involve NRC regulated activities.

Consideration must be given to all information collected in making a T&R determination for unescorted access. Licensees must substantiate and document potentially disqualifying information obtained from confidential or unnamed sources and should not use this information as the sole basis to deny access authorization unless corroborated. Licensees should establish criteria that would disqualify someone from being granted authorized access. In every case, the licensee should evaluate T&R based on an accumulation of information that supports a finding with reasonable assurance. The licensee is responsible for making the T&R determination for an employee who is seeking unescorted access. The T&R determination is designed to identify past actions pertinent to whether reasonable assurance of an individual's future reliability exists. In addition to the criminal history records check, licensees may want to consider if the subject individual has exhibited the following behaviors:

- impaired performance attributable to psychological or other disorders
- conduct that warrants referral for criminal investigation or results in an arrest or conviction
- an indication of deceitful or delinquent behavior
- attempted or threatened destruction of property or life
- suicidal tendencies or an attempt at suicide
- illegal drug use or the abuse of legal drugs
- alcohol abuse disorders

- recurring financial irresponsibility
- irresponsibility in the performance of assigned duties
- inability to deal with stress or the appearance of being under unusual stress
- failure to comply with work directives
- hostility or aggression toward fellow workers or authority
- uncontrolled anger, violation of safety or security procedures, or repeated absenteeism
- significant behavioral changes, moodiness, or depression

These indicators are not meant to be all inclusive or intended to be disqualifying factors. Licensees also may consider extenuating or mitigating factors in their determinations.

ANNEX B. SAMPLE CONSENT FORM FOR BACKGROUND INVESTIGATIONS

I authorize and grant my consent to my employer, ________ *(insert company name)*_______ _______ (hereinafter or "the company"),________ to request the U.S. Nuclear Regulatory Commission (NRC), under Section 652 of the Energy Policy Act of 2005, to request criminal record information about me from the U.S. Attorney General, who will refer the request to the Federal Bureau of Investigation. I understand that the purpose of this information is solely to enable the company to determine my trustworthiness and reliability for unescorted access to a Category 2 or greater quantity of radioactive material as defined in the NRC's regulations in Title 10 of the *Code of Federal Regulations* (10 CFR) Part 37, "Physical Protection of Category 1 and Category 2 Quantities of Radioactive Material." I understand that ________ *(insert company name)*_______ must obtain my signed consent before any investigation or reinvestigation to determine my trustworthiness and reliability for such unescorted access.

I authorize and grant my consent to any authorized representative of ________ *(insert company name)*_______ who is conducting my background investigation or reinvestigation, as defined in 10 CFR 37.25, "Background Investigations," to obtain any information related to my activities from individuals, schools, residential management agents, previous employers, criminal justice agencies, or other sources of information. This information may include, but is not limited to, my academic, residential, achievement, or performance information and information about my attendance, disciplinary, employment, and criminal history records.

I authorize the Federal Bureau of Investigation to disclose the record of my criminal history background investigation to my employer for the purpose of making a determination of my trustworthiness and reliability for unescorted access to a category 2 or greater quantity of radioactive material. I understand that before making any determination to deny me this unescorted access, ________ *(insert company name)*_______ will provide me a copy of the information on which it intends to base that determination. I further understand that before a final adverse determination, my employer must give me an opportunity to correct any inaccurate or incomplete information that is developed during the background investigation.

I understand that I may withdraw my consent at any time and that after I do withdraw my consent, under 10 CFR 37.23(c) of the NRC's regulations, ________ *(insert company name)*________ may not initiate any elements of the background investigation that were not in progress at the time that I withdrew my consent. I also understand that, under 10 CFR 37.23(c), the withdrawal of my consent for the background investigation is sufficient cause for denial or termination of any authorization for unescorted access.

I understand that, for previous employers and other sources of information, separate specific releases may be needed and that I may be contacted for such releases at a later date. I authorize custodians of records and other sources of information pertaining to me to release such information upon request of the investigator or other duly authorized representative of the company regardless of any previous agreement to the contrary.

I understand that the information released by records custodians and other sources of information is solely for the purpose of making a determination about my trustworthiness and reliability for unescorted access to the radioactive materials subject to 10 CFR Part 37 and that this information may be disclosed only as authorized by State or Federal law.

I understand that photocopies of this authorization and consent document with my signature are valid and that this authorization will remain in effect as long as I am authorized to obtain unescorted access to the radioactive material subject to 10 CFR Part 37.

Signature (In Ink)	Full Name (Type or Print Legibly)			Date Signed
Other Names Used		Date of Birth		Social Security No.
Current Street Address and Apartment No.	City and County	State	Zip Code	Home Telephone No.

Subpart C — Physical Protection Requirements during Use

§ 37.41, "Security Program"
§ 37.43, "General Security Program Requirements"
§ 37.45, "LLEA Coordination and Notification"
§ 37.47, "Security Zones"
§ 37.49, "Monitoring, Detection, and Assessment"
§ 37.51, "Maintenance and Testing"
§ 37.53, "Requirements for Mobile Devices"
§ 37.55, "Security Program Review"
§ 37.57, "Reporting of Events"

§ 37.41, "Security Program"

§ 37.41(a), "Applicability"

§ 37.41(a)(1)

Each licensee that possesses an aggregated category 1 or category 2 quantity of radioactive material shall establish, implement, and maintain a security program in accordance with the requirements of this subpart.

Explanation

Each licensee that possesses an aggregated quantity of category 1 or category 2 quantities of radioactive material must have a security program.

Q&As

Q1: What is the purpose of a security program?

A1: The purpose of a security program under 10 CFR Part 37 is to plan and document how the licensee will protect category 1 and 2 quantities of radioactive material. The program specifies how features, such as training, access controls, detection and assessment methods, and response capabilities, will function together to mitigate the potential detrimental consequences to public health and safety and the environment.

Q2: What does it mean to have an "aggregated" quantity of radioactive material?

A2: The NRC considers radioactive material to be "aggregated" if someone could gain access to a category 2 or greater quantity by breaching a single physical barrier. "Aggregated" has the same meaning as "collocated" in the guidance for the IC orders. The regulations in 10 CFR Part 37 use the term "aggregated" instead of "collocated" to avoid the confusion that could arise when several separate non-aggregated quantities of radioactive material are located at the same site or inside the same facility.

Q3: Does 10 CFR Part 37 apply if I have two or more sources at the same location that, when added together, meet or exceed a category 2 threshold quantity?

A3: The regulations in 10 CFR Part 37 apply when an additional physical barrier between the sources or devices is not present. In such a case, an intruder would only have to defeat one barrier to gain access to a category 2 or greater quantity of radioactive material. With only one common barrier, the sources or devices are considered aggregated, and the licensee will have to develop and implement a security program. An example would be a high-dose afterloader with a backup source. Two sources stored in the same area behind a single barrier would be aggregated, and 10 CFR Part 37 would apply. In this same example of a high-dose afterloader and backup source, 10 CFR Part 37 would not apply if the licensee stores the backup source in a separate locked room or in a locked container that could not be removed, such as a safe. These would be considered additional barriers, and the afterloader and backup source would no longer meet the definition of "aggregated" because they would no longer be isolated by a single physical barrier.

Q4: How do I determine whether I possess a category 1 or category 2 quantity of radioactive material?

A4: The licensee should use the sum-of-fractions method, also known as the unity rule, to determine if it possess a category 1 or category 2 quantity of radioactive material. The licensee may need to implement 10 CFR Part 37 requirements even if it does not possess any single source or single radionuclide in excess of a category 2 threshold. For combinations of materials (including sealed sources, unsealed sources, and bulk material) and radionuclides, a licensee must include multiple sources (including bulk material) of the same radionuclide and multiple sources (including bulk material) of different radionuclides to determine if the requirements apply. For the purposes of this calculation, licensees must consider all the radioactive material they possess under their licenses at the location in question.

The licensee can use the following formula for the unity rule to determine if it has a category 2 quantity and if it is required to implement 10 CFR Part 37 requirements:

$$R1/AR1 + R2/AR2 + R_n/AR_n > 1.0$$

where:

R1 = total amount of radionuclide 1
AR1 = category 2 threshold of radionuclide 1
R2 = total amount of radionuclide 2
AR2 = category 2 threshold of radionuclide 2 R_n = total amount of radionuclide n
AR2 = category 2 threshold of radionuclide n and so on greater than or equal to 1.0

The licensee would use the same unity rule formula above to determine if it has a category 1 quantity for purposes of implementing 10 CFR Part 37 requirements. It would need to substitute, however, the category 1 threshold value as the divisor for each radionuclide. The category 1 thresholds in Appendix A are 100 times the thresholds for category 2 quantities.

The licensee must use terabecquerel thresholds in all unity rule calculations because terabecquerel is the regulatory standard. Therefore, the licensee must use the following formula to convert any curie values in its license to terabecquerel:

$$n\ (\text{TBq}) = N\ (\text{Ci}) \times 0.037\ \text{TBq/Ci}$$

See the Q&As for the definitions of category 1 and category 2 quantities in 10 CFR 37.5 for illustrative examples of the unity rule and terabecquerel conversion calculations.

Q5: If I have radioactive materials aggregated to a category 2 or greater quantity, what do I have to do to meet the separate requirements of 10 CFR 37.41(a)(1) to "establish," to "implement," and to "maintain" a security program?

A5: To "establish" the security program, the licensee must develop a written security plan that describes a strategy and a set of technologies and must develop measures that, when implemented with written procedures that are also required, will demonstrate compliance with each of the applicable requirements in Subpart C of this rule. Under § 37.41(c) below, the security program must, as appropriate, include the program features described in 10 CFR 37.43, "General Security Program Requirements"; 10 CFR Parts 37.45; 37.47; 37.49; 37.51; 37.53; and 37.55, "Security Program Review." The specific measures that must be implemented under the security program should include, among other things, the establishment of security zones; the design and procurement of a physical protection system for monitoring, detection, assessment, and alarm/communication; the procurement of any necessary support services, such as a commercial security service between normal working hours; and coordination with affected LLEAs. The program must also include recordkeeping measures and must provide for the operation, testing, and maintenance of equipment and technologies that, when functioning as an integrated system with the other measures required under this section, are designed to effectively monitor and, without delay, detect, assess, and respond to an actual or attempted unauthorized access to category 1 or category 2 quantities of radioactive material.

To "implement" the security program, the licensee must have trained necessary staff under 10 CFR 37.43(c) to follow its program procedures; must have installed and must be using all the access control measures or systems required under 10 CFR 37.43(a)(1)(ii); must have completed initial LLEA coordination efforts under 10 CFR 37.45; must have installed the physical barriers protecting each security zone in accordance with 10 CFR 37.47; and must have installed and must be using all the monitoring, detection, and alarm/communications systems required under 10 CFR 37.49. The licensee must also have placed into service all the appropriate other measures, such as those for testing and maintenance under 10 CFR 37.51; those for mobile devices under 10 CFR 37.53; those for reporting of events under 10 CFR 37.55; and those for program reviews under 10 CFR 37.57, "Reporting of Events," that are required by this part and that the licensee will rely upon to meet the performance objectives of 10 CFR 37.41(b).

To "maintain" the security program, the licensee must show that it has implemented and is continuing to fund the program's access authorization and control, training, monitoring, detection, assessment, alarm and communication, and LLEA coordination activities and to inspect, test, and repair the equipment installed to keep it functioning properly. The licensee will also need to periodically evaluate and update operating, training, and LLEA coordination procedures and other measures required by this part, as appropriate, to address lessons learned and to ensure the continuing compliance of the system that it has designed and implemented to meet the performance objectives of 10 CFR 37.41(b).

Q6: Must I have a security program for temporary jobsites?

A6: Yes. If a licensee has a category 1 or category 2 quantity of radioactive material at a temporary jobsite, it must have a security program that meets the requirements of 10 CFR Part 37. However, this does not mean that the licensee must develop and implement a security program specifically for each temporary jobsite. It only means that the licensee's security plan under 10 CFR 37.43(a) must address the security of its category 1 or category 2 quantities of radioactive material while working at temporary and permanent jobsites. Additionally, when transporting radioactive material in such quantities to and from a temporary jobsite, the licensee must control access and must monitor, detect, assess, and respond to an actual or attempted theft, sabotage, or diversion. These measures against unauthorized access must also be applied when the transport vehicle is stopped at a hotel, restaurant, gas station, or other location. (See the Q&As on 10 CFR 37.79(a)(2).)

Q7: If I'm a licensed contract radiographer, and I must leave a radiography camera at another licensee's or customer's facility that provides its own site security, who is responsible for security? When I'm expected to leave a camera on site at an oil refinery, for example, who provides security there, and if the customer provides the security, how can I know whether the security satisfies 10 CFR Part 37?

A7: A licensee that possess radioactive material is responsible for its security and control, and the licensee must meet 10 CFR Part 37 security requirements in its own facility and in a customer's facility. If the licensee chooses to store one of its devices at a customer's facility, the licensee and the customer must ensure that they meet the applicable 10 CFR Part 37 requirements. The licensee will need to evaluate if its customer's physical protection program will comply with the applicable 10 CFR Part 37 requirements at the agreed storage location. In addition, the licensee and the customer should have a clear understanding of the licensee's respective roles and responsibilities and the features of the customer's security and control program that must be relied upon to meet the applicable security requirements for the licensee's device. The licensee should assess the customer's security systems and measures against those of its own program for implementing these requirements.

§ 37.41, "Security Program" (continued)

§ 37.41(a), "Applicability"

§ 37.41(a)(2)

An applicant for a new license and each licensee that would become newly subject to the requirements of this subpart upon application for modification of its license shall implement the requirements of this subpart, as appropriate, before taking possession of an aggregated category 1 or category 2 quantity of radioactive material.

§ 37.41(a)(3)

Any licensee that has not previously implemented the security orders or been subject to the provisions of Subpart C shall provide written notification to the NRC regional office specified in 10 CFR 30.6 of this chapter at least 90 days before aggregating radioactive material to a quantity that equals or exceeds the category 2 threshold.

Explanation

New license applicants and licensees newly subject to 10 CFR Part 37 must implement the requirements of Subpart C before taking possession of an aggregated category 1 or category 2 quantity of radioactive material. A licensee that has not previously implemented the security orders or had not been subject to the provisions of Subpart C must notify the appropriate NRC regional office in writing at least 90 days before aggregating radioactive material to a quantity that equals or exceeds the category 2 threshold.

Q&As

Q1: At what point in the process of transferring possession of radioactive material would I be considered to have "taken possession" of a category 1 or category 2 quantity? Have I "taken possession" of a source under 10 CFR Part 37 when the carrier or delivery service crosses my site boundary, for example? When the carrier has brought the source into my facility building? When the source is in a permanent security zone on my site?

A1: The licensee is considered by the NRC to have taken possession of a category 1 or category 2 quantity when it has accepted the consignment of the material from the shipper or transferring licensee and has assumed physical control of the material. Thus, the licensee is considered by the NRC *not* to have taken possession when the shipper has crossed its site boundary or has brought the source into a building on its facility site unless it accepts the consignment and assumes physical control of the material at that point on its premises. At that point, if accepting the consignment causes the licensee to have aggregated radioactive material in a category 2 or greater quantity, it must meet the requirements in 10 CFR Part 37.

If the licensee orders new byproduct material that would put it over the category 2 limit for an aggregated quantity when it arrives at its facility, it should have established its security program before it ordered the material, including the access authorization program required by Subpart B.

Q2: Would I be considered to have "taken possession" if the carrier delivers a shipment and then leaves without my having accepted the consignment?

A2: Yes. The licensee would be held accountable if the carrier left a shipment on the doorstep or loading dock. The licensee is responsible for making arrangements for receipt of the shipment. However, this scenario, except in extreme or emergency circumstances, is highly unlikely. The shipping licensee is required, under 10 CFR 37.79(a)(3)(iii), to use a carrier that requires a signature for receipt of category 2 shipments.

Q3: If I am receiving multiple deliveries of radioactive material, when would I be considered to have taken possession of an "aggregated" category 1 or category 2 quantity?

A3: The NRC would consider the licensee to have taken possession of an "aggregated" category 1 or category 2 quantity of radioactive material when and if it places an accumulated quantity at or above the category 2 threshold behind a single physical barrier. If the licensee places one or more of these deliveries in a total quantity that is less than category 2 behind a separate, independent physical barrier, it is not taking possession of an "aggregated" quantity of material as defined in 10 CFR Part 37.

§ 37.41, "Security Program" (continued)

§ 37.41(b), "General Performance Objective"

Each licensee shall establish, implement, and maintain a security program that is designed to monitor and, without delay, detect, assess, and respond to an actual or attempted unauthorized access to category 1 or category 2 quantities of radioactive material.

§ 37.41(c), "Program Features"

Each licensee's security program must include the program features, as appropriate, described in §§ 37.43, 37.45, 37.47, 37.49, 37.51, 37.53, and 37.55.

Explanation

This provision establishes the general performance objective of the security program and specifies the features it must include, as appropriate.

Q&As

Q1: What does it mean for a licensee's security program to be designed to detect, assess, and respond to an unauthorized access event "without delay"?

A1: The licensee must have a security program that enables it to detect unauthorized access to the security zone when it occurs, to determine whether the unauthorized access was an actual or attempted theft, and to initiate an appropriate response right away. The objectives are to reduce the risk that the material will be stolen and used for an unauthorized purpose and, if it is stolen, to improve the likelihood of timely recovery or to mitigate potential consequences.

Q2: What's the difference between "actual" and "attempted" unauthorized access? Wouldn't the same kinds of measures apply to both?

A2: Yes. The same kinds of measures apply to both. Although the measures for immediate detection, assessment, and response to *actual* unauthorized access to radioactive materials also will be required for *attempts* to gain unauthorized access, licensees should be alert for covert actions in preparation for an attempted theft, sabotage, or diversion. For example, such covert actions could include the theft and copying of keys to locked rooms or storage containers or the gaining of access to security codes for copying onto duplicate key cards. Licensees should therefore monitor, assess, and respond to actual or attempted unauthorized access to keys, security cards, codes, or other means that could be used in an attempt to gain unauthorized access to the material in a security zone.

Q3: Can licensees use measures required for radiation safety to comply with 10 CFR Part 37 security requirements too?

A3: Yes. The licensee may use or modify systems used to control access to high-radiation areas, as required by 10 CFR Part 20, "Standards for Protection against Radiation," or equivalent Agreement State regulations, or other detection and access control systems used for radiation protection for security purposes as long as the uses or modifications do not compromise the original safety purpose of these systems. Documentation should describe how these systems provide the required intrusion detection.

Q4: What are the principal requirements of a security program?

A4: The licensee must have a security program to continuously monitor and to immediately detect unauthorized access to category 1 or category 2 quantities of radioactive material when the unauthorized access occurs. The program also must enable the licensee to determine if the unauthorized access is an actual or attempted theft and to initiate an appropriate response without delay. The detection system must be capable of detecting all unauthorized access to the security zone, including breaches of barriers used to isolate and control access to the protected radioactive material. It must also be capable of detecting an unauthorized removal of protected material from security zones. Assessment can be done either by automated devices or trained personnel who can initiate the appropriate response. Licensees should consider the possibility of simultaneous alarms at multiple locations. The program's documentation must also describe the processes that the licensee will use to assess and respond to unauthorized access.

A licensee's security program must include a written security plan, implementing procedures, training, the use of security zones, the protection of information, requests for coordination with the cognizant LLEA(s), additional security measures for mobile devices (if applicable), testing and maintenance of security-related equipment, an annual program review, and incident reporting requirements. Each of these areas is discussed in more detail in the following Q&As.

§ 37.43, "General Security Program Requirements"

§ 37.43(a), "Security Plan"

§ 37.43(a)(1)

Each licensee identified in § 37.41(a) shall develop a written security plan specific to its facilities and operations. The purpose of the security plan is to establish the licensee's overall security strategy to ensure the integrated and effective functioning of the security program required by this subpart. The security plan must, at a minimum:

§ 37.43(a)(1)(i)

Describe the measures and strategies used to implement the requirements of this subpart; and

§ 37.43(a)(1)(ii)

Identify the security resources, equipment, and technology used to satisfy the requirements of this subpart.

Explanation

Subject licensees must develop a written security plan with an overall security strategy to ensure the integrated and effective functioning of the security program. This subsection also specifies minimum contents of the security program.

Q&As

Q1: Must licensees submit their written security plans to the NRC?

A1: No. The regulations in 10 CFR Part 37 do not require, and the NRC does not want, licensees to submit security plans or implementing procedures to the agency for review or approval. The security plan will be subject to NRC inspection.

Q2: What must a licensee's security plan address?

A2: As noted in 10 CFR 37.43(a)(1), the plan must, among other things, include a description of the measures and strategies to implement the security requirements and to identify the resources, equipment, and technology used. As good practice in explaining its overall strategy, the licensee should ensure that its security plan describes any site-specific conditions that affect how it will implement the requirements. Security plans are important for the implementation of a performance-based regulation. An adequate plan requires a licensee to analyze the particular security needs of its individual facilities and to explain clearly how it will implement its chosen security measures to ensure that they work together to meet the applicable performance objectives.

To ensure the integrated and effective functioning of the security program and to facilitate its meeting the program review requirements in 10 CFR 37.33, the licensee also may consider describing, in the security plan, a process for identifying and implementing corrective actions or compensatory measures in the event of a failure of personnel or equipment to perform as specified or function as required.

One method for developing a security plan would be to structure the plan in accordance with Appendix II to IAEA's implementing guide entitled, "Security of Radioactive Sources," IAEA Nuclear Security Series No. 11, issued May 2009 (http://wwwpub.iaea.org/MTCD/publications/PDF/Pub1387 web.pdf). Although Appendix II to the guide notes that "[t]he level of detail and depth of content [of a licensee's security plan] should be commensurate with the security level of the source(s) covered by the plan" and although not all of the recommended measures may apply to a given licensee's situation, the appendix establishes the recommended contents of a typical security plan for the radioactive materials subject to 10 CFR Part 37. For easy reference, the following list provides these contents:

- a description of the radioactive material, its categorization, and its use
- a description of the environment, building, and facility in which the radioactive material is used or stored and, if appropriate, a diagram of the facility layout and security system
- the location of the building or facility relative to areas accessible to the public
- local security procedures
- objectives of the security plan for the specific building or facility, including the following:
    - the specific concern that will be addressed (e.g., unauthorized removal, destruction, or malevolent use)
    - the kind of control necessary to prevent undesired consequences, including the auxiliary equipment that might be needed
    - the equipment or premises that will be secured
- security measures that will be used, including the following:
    - the measures to secure, provide surveillance, provide access control, detect, delay, respond, and communicate
    - the design features to evaluate the quality of the measures against the assumed threat
- administrative measures that will be used, including the following:
    - security roles and responsibilities of management, staff, and others
    - routine and non-routine operations, including an accounting of the source(s)
    - maintenance and testing of equipment
    - a determination of the trustworthiness of personnel
    - the application of information security
    - methods for access authorization
    - security related aspects of the emergency plan, including event reporting
    - training
    - key control procedures
    - procedures to address an increased threat level
    - the process for periodically evaluating the effectiveness of the plan and updating it accordingly
    - any compensatory measures that may need to be used
    - references to existing regulations or standards

Q3: Can the security plan and implementing procedures be in the same document?

A3: Yes. However, any document or group of documents that contain both the security plan and its implementing procedures must meet the respective requirements in 10 CFR 37.41(a) and 10 CFR 37.41(b) for each of these program elements.

Q4: Is a licensee required to develop specific additional contingency plans for situations where a facility or site needs to evacuate staff due to an emergency, natural disaster, or other events where public health and safety are threatened?

A4: The regulations in 10 CFR Part 37 do not specifically require a licensee to develop any contingency plans as such; however, contingency planning is considered a best practice in the development of security plans. The NRC encourages licensees to develop contingency planning as consistent with the intent of 10 CFR Part 37 and the security culture it seeks to promote. Because licensees are required to ensure the security and accountability of the radioactive material protected under this part, they should also consider measures to maintain this security and accountability in the event of an evacuation or other emergency situation. If developed, these plans should address things, such as floods, earthquakes, and tornadoes that have an increased probability of occurring in the area in which the facility or site is located.

Q5: Should the radiation safety office or officer be involved in the development or implementation of security plans and procedures?

A5: Although RSOs and their staff may not be the licensee's experts in security, it should involve them because they can provide valuable insights for improving the consideration of safety and security risks in a system integrated way. The RSO or officer can also provide advice and analysis to ensure that the licensee is implementing security requirements in a manner that does not compromise safety.

Q6: Can a licensee with a security plan already in place for the increased controls orders be automatically exempted from requirements to develop and implement a new security plan under Part 37?

A6: No. The requirements in 10 CFR Part 37 do not automatically exempt the licensee from having a security plan. If the licensee has an existing security plan, it will need to check that the plan meets all the requirements for a security plan in 10 CFR Part 37. If it does meet the requirements, the licensee may continue to use its existing plan.

Q7: If I already have a high level corporate security plan in place that meets the requirements of 10 CFR 37.43(a), do I still have to develop and implement a separate security plan to demonstrate compliance with 10 CFR Part 37?

A7: No. As long as a preexisting site or corporate-wide plan meets the requirements of Subpart C in regard to the content of the security plan, the NRC would consider the plan acceptable, and the licensee would not need to develop a new plan. Each site would need to have a copy of the plan for implementation.

Q8: If I'm a licensee possessing a category 2 or greater quantity of radioactive material and I hire a service contractor licensed by the NRC or another Agreement State, do the requirements of the contractor's security program govern the contractor employee when he or she is on my site, or must the employee comply with the requirements of my security program?

A8: The answer to this question depends on the ownership of the material. If the licensee owns the radioactive material or the devices that contain the material, such as in the case in which a hospital owns a gamma stereotactic radiosurgery device, its contractor would be governed by its security program. If the contractor owns the material or device, such as a radiography camera that the contractor brings to the licensee's site, the contractor's security requirements would apply to the use of that material or device and its control.

§ 37.43, "General Security Program Requirements" (continued)

§ 37.43(a), "Security Plan"

§ 37.43(a)(2)
The security plan must be reviewed and approved by the individual with overall responsibility for the security program.

Explanation

This definition is self-explanatory. Q&As:

Q1: Who should be considered "the individual with overall responsibility for the security program"?
A1: The licensee can designate the individual. The individual can be the company president, chief executive, the RSO, or any other individual who has been designated as the responsible person for the security program. The individual responsible for the security of all the licensee's category 1 or category 2 quantities of radioactive material is the appropriate individual to provide written approval of the plan.

§ 37.43, "General Security Program Requirements" (continued)

§ 37.43(a), "Security Plan"

§ 37.43(a)(3)
A licensee shall revise its security plan as necessary to ensure the effective implementation of Commission requirements. The licensee shall ensure that:

§ 37.43(a)(3)(i)
The revision has been reviewed and approved by the individual with overall responsibility for the security program; and

§ 37.43(a)(3)(ii)
The affected individuals are instructed on the revised plan before the changes are implemented.

Explanation

A licensee must revise its security plan as needed to ensure effective implementation. The individual with overall responsibility for the security program must review and approve any revision, and the licensee must instruct the affected individuals on the revised plan before it implements the changes.

Q&As

Q1: How frequently should I revise my security plan?

A1: No predetermined frequency exists for the revision of the security plan. The licensee may want to revise its plan after the completion of the annual security program review so that the revised plan addresses any recommendations. The licensee may need to revise its security plan (1) if it increases the quantity of radioactive material that it has aggregated at a given location, (2) if it moves the location of a storage area for an aggregated quantity, or (3) if it alters its facility in a way that could affect the security of its licensed radioactive material subject to this part. In addition, the licensee would need to revise its security plan before it changes the measures that it relies upon to comply with 10 CFR Part 37. For example, if the licensee plans to shift from direct surveillance to an offsite monitored alarm system, it would need to address that change in revisions to its security plan and procedures.

Q2: Who should approve revisions to the security plan under 10 CFR 37.43(a)(3)(i)?

A2: The individual who has overall responsibility for the security program must approve the revised security plan. As noted in the discussion of 10 CFR 37.43(a)(2), the approving individual may be the company president, chief executive, the RSO, or any other individual who has been given the responsibility for the security program.

Q3: 10 CFR 37.43(a)(3)(ii) requires me to ensure that "affected individuals" are "instructed" on the revised plan before the changes are implemented. How broadly or narrowly should I define "affected individuals"? Do I have to retrain all my employees and any service contractor employees before I can put any change in my security plan into effect?

A3: No. The licensee does not need to retrain all its employees or contractors before implementing a change in its security plan. The licensee must retrain only those employees whose responsibilities or job duties would be impacted.

Q4: 10 CFR 37.43(a)(3)(ii) also requires me to ensure that these affected individuals are "instructed" on the revised plan before the changes are implemented. What must I require of the "affected individuals" to show that they have been sufficiently "instructed"? Must all the affected individuals get a passing score on a proficiency exam or demonstrate the needed new skills before I can put the change into effect?

A4: No. The licensee does not need to require all the affected individuals to pass a proficiency exam or demonstrate mastery of the necessary skills before implementing a change in its security plan. Depending on the change, the instruction could include the reading of new procedures, a briefing on the new item(s), or on-the-job training. As with any training, the licensee needs to document that the training occurred. The licensee should consider the scope of the security plan revision in relation to the scope of each individual's job duties to identify those individuals who may need to show a more measurable level of knowledge or skill to ensure the effective or timely implementation of the change in its security plan. A change that would affect the licensee's policy on employee observation or on the reporting of suspicious activities may only require a briefing before its implementation. As a good practice to demonstrate compliance with this instruction requirement, the licensee should require the affected individual(s) to sign a statement that he or she has read, and is familiar with, the relevant portion of the security plan and procedures.

§ 37.43. "General Security Program Requirements" (continued)

§ 37.43(a), "Security Plan"

§ 37.43(a)(4)

The licensee shall retain a copy of the current security plan as a record for 3 years after the security plan is no longer required. If any portion of the plan is superseded, the licensee shall retain the superseded material for 3 years after the record is superseded.

Explanation

This definition is self-explanatory.

Q&As

Q1: If I've received a license amendment that reduces my possession limits to quantities below a category 2 threshold, and my prelicense amendment records are inspected a few months or a year later, am I still required to maintain these records for the rest of the 3-year period?

A1: Yes. The licensee must retain most records, including those preceding its license amendment, for 3 years. However, it must maintain some records until its license is terminated. In the case of the security plan, the licensee must retain the record for 3 years after the plan is no longer needed. The licensee may destroy all records related to 10 CFR Part 37 once its license has been terminated.

Q2: What does it mean for the security plan or any portion of it to be "superseded"?

A2: The NRC would consider the security plan to be "superseded" once a revision to the plan has been approved. To be considered superseded, a plan can be completely revised or partially revised. The licensee would need to retain a copy of the old security plan for 3 years or until license termination, if that occurs first.

§ 37.43, "General Security Program Requirements" (continued)

§ 37.43(b), "Implementing Procedures"

§ 37.43(b)(1)

The licensee shall develop and maintain written procedures that document how the requirements of this subpart and the security plan will be met.

§ 37.43(b)(2)

The implementing procedures and revisions to these procedures must be approved in writing by the individual with overall responsibility for the security program.

§ 37.43(b)(3)

The licensee shall retain a copy of the current procedure as a record for 3 years after the procedure is no longer needed. Superseded portions of the procedure must be retained for 3 years after the record is superseded.

Explanation

Subject licensees must have written procedures that document how they will implement the security plan and other Subpart C requirements. The individual with overall responsibility for the program must approve, in writing, procedures and future revisions. The licensee must keep a copy of the current procedure as a record for 3 years after the procedure is no longer required. If any portion is superseded, the licensee must keep the superseded material for 3 years.

Q&As

Q1: Am I required to develop security procedures if I don't actually possess a category 2 quantity of material?

A1: No. A licensee only needs to develop security procedures if it possesses an aggregated quantity that meets or exceeds the category 2 threshold and must therefore implement a security program. If a licensee plans on obtaining additional radioactive material or on relocating currently possessed radioactive material into an aggregated quantity that meets or exceeds the category 2 threshold, it must establish its security program before it can obtain or relocate the material. This would include having the procedures in place.

Q2: What must a licensee's security procedures address?

A2: Generally, the security procedures must address how the licensee will implement the applicable features required by Subpart C. Depending on the licensee and its operating requirements, these features would require its procedures (1) to address training, (2) to establish and maintain security zones, and (3) to establish the monitoring, detection, assessment, and response measures; maintenance and testing measures; the reporting of events; and the periodic review of the program.

The written procedures should address how the licensee will respond to a range of foreseeable events common to the type of license. Examples of such events could range from an inadvertent unauthorized access that would not require an LLEA response to a malevolent intrusion that would require LLEA intervention. These procedures should include, if applicable, provisions for immediate response, for after hour notification of LLEAs and the licensee individual responsible for security, for the handling of both radiation safety and security related types of emergencies, and for events at temporary jobsites.

Procedures should also address the roles of the licensee's staff and, where applicable, its contractors. The licensee's staff and contractors must have a clear understanding of their responsibilities and constraints in an emergency, along with step-by-step instructions and clear guidelines for whom to contact. However, note that when developing these security procedures, the licensee must not compromise facility operational safety, occupational safety, fire safety, and emergency planning at the facility.

Q3: Is there flexibility built into the 10 CFR Part 37 implementation process?

A3: Yes. The regulations in 10 CFR Part 37 are purposely not prescriptive, which allows licensees to tailor programs to their specific facility and operations. Various approaches are available for licensees to meet the objectives of10 CFR Part 37, and no one solution exists for any material control challenge facing licensees. This guidance document provides examples of how licensees may meet 10 CFR Part 37 requirements. Licensees do not have to implement any of the examples in the guidance; each example describes only one acceptable method that licensees can use to comply with 10 CFR Part 37 requirements.

Q4: Who can approve implementing procedures?

A4: The individual considered to have overall responsibility for the licensee's security program is the individual who must approve any security implementing procedures. (See 10 CFR 37.43(a)(2).)

Q5: What does it mean for the security procedure or any portion of it to be "superseded"?

A5: A procedure is superseded when a new version of the procedure has been issued. A procedure can be completely revised or partially revised to be considered superseded. The licensee would need to retain a copy of the old procedure for 3 years or until license termination if that occurs before then.

§ 37.43, "General Security Program Requirements" (continued)

§ 37.43(c), "Training"

§ 37.43(c)(1)

Each licensee shall conduct training to ensure that those individuals implementing the security program possess and maintain the knowledge, skills, and abilities to carry out their assigned duties and responsibilities effectively. The training must include instruction in:

§ 37.43(c)(1)(i)

The licensee's security program and procedures to secure category 1 or category 2 quantities of radioactive material and in the purposes and functions of the security measures employed;

§ 37.43(c)(1)(ii)

The responsibility to report promptly to the licensee any condition that causes or may cause a violation of Commission requirements;

§ 37.43(c)(1)(iii)

The responsibility of the licensee to report promptly to the local law enforcement agency and licensee any actual or attempted theft, sabotage, or diversion of category 1 or category 2 quantities of radioactive material; and

§ 37.43(c)(1)(iv)

The appropriate response to security alarms.

§ 37.43(c)(2)

In determining those individuals who shall be trained on the security program, the licensee shall consider each individual's assigned activities during authorized use and response to potential situations involving actual or attempted theft, diversion, or sabotage of category 1 or category 2 quantities of radioactive material. The extent of the training must be commensurate with the individual's potential involvement in the security of category 1 or category 2 quantities of radioactive material.

Explanation

Individuals implementing the security program must be trained. The training must cover the program and its procedures, the purpose and function of the security measures, and the individual's responsibilities. The extent of the training must be commensurate with the individual's potential involvement in radioactive material security.

Q&As

Q1: What is the purpose of the training program?

A1: The purpose of the training program is to ensure that those employees who are responsible for implementing security measures know what is expected of them and know how to do their job.

Q2: What minimum elements must a training program include?

A2: A licensee's training program should be evaluated against the Subpart C performance objective for ensuring that each individual with responsibility for any aspect of the security program has the requisite knowledge, skills, and abilities to carry out that responsibility effectively. Thus, the minimum scope of a training program should identify these individuals and identify the working knowledge, skill sets, and capabilities they need to carry out their assigned duties and responsibilities for effective implementation of the licensee's security plan. Examples of appropriate subjects for training would include the controls that are in place to prevent unauthorized access to material, the purpose and functional requirements of the licensee's alarm and access control systems, notification procedures in the event of an unauthorized access for potential malevolent activities, and ways to confirm quickly and accurately if an intrusion is likely to be intentional or accidental. In addition, an adequate program must train employees to identify any "condition[s] that cause...or may cause a violation" of Commission requirements and to identify and report suspicious activities. An adequate program should also cover the operation of primary and backup communication systems for such reporting.

The licensee's staff members must have a clear understanding of their individual responsibilities and constraints in an emergency, and training should provide step-by-step instructions and clear guidelines for what to do and for whom to contact. In addition, the training should provide step-by-step instructions and clear guidelines on the proper performance of testing and maintenance activities. Training does not need to be in a classroom and can be performed as a part of on-the-job training. Training may be organized by subject area as well. For example, the licensee might train most of its employees on escort responsibilities and general alarm response. The licensee should train those employees who will be conducting surveillance on what to look for and how to respond when they see something unusual.

Q3: What "potential situations" involving theft, diversion, or sabotage should a licensee consider when determining who should be trained?

A3: A single set of potential security situations that would apply to every licensee's individual security needs does not exist. Each licensee's security plan should develop and evaluate a set of site-specific theft, diversion, and sabotage scenarios that could arise from the licensee's particular operating requirements and worksite settings with sufficient probability, consequences, or both to make it prudent for the licensee to address. The training program should address these scenarios. The licensee must evaluation potential situations for the monitoring, detection, assessment, and response functions needed for a timely and effective response; scenarios could include events ranging from an inadvertent unauthorized access that would not require an LLEA response to a malevolent intrusion that would. These procedures should include provisions for immediate response, for after hour notification of LLEAs and licensee management, and for the handling of credible radiation safety problems that could result from postulated security emergencies. Where applicable, the licensee should also develop training requirements for security events at temporary jobsites.

Q4: The rule requires the extent of the licensee's security training for affected individuals to be "commensurate with the individual's potential involvement" in the subject material's security. Should all individuals with a valid reason to be on site be considered to have "potential involvement" in a security emergency and, therefore, require training?

A4: No. Although it is good practice to provide some basic training to all employees (e.g., details on how to respond to an alarm), only those individuals with security responsibilities must be trained. If an employee does not have unescorted access to the radioactive material, does not have access to any security information, and has no security duties, he or she may not need to receive any security training beyond the basics of safely responding to alarms. In addition, not all individuals need to receive the same level of training.

Q5: How should a licensee demonstrate that the individuals responsible for implementing the security plan have acquired the necessary "knowledge, skills, and abilities" to carry out their assigned duties and responsibilities "effectively"?

A5: Although testing is not required, one measure of the likelihood that a trainee will be able to carry out assigned security responsibilities is evidence that the trainee took and passed a reasonable test of the knowledge, skill, or ability objectives of the training. Another measure of the likelihood that an employee or contractor will be able to carry out assigned security responsibilities is evidence that he or she has successfully participated in a drill or exercise designed to test the integrated functionality of the entire security system, including monitoring, detection, assessment, and response. Using such system-wide drills or tabletop exercises, and notifying affected LLEAs of opportunities to participate in such training, can be an especially effective way for some licensees to train their personnel.

However, depending on the frequency and relative complexity of an individual's assigned duties, adequate training can also be demonstrated through the daily or weekly conduct of that individual's job. This could be especially true for some elements of refresher training. For example, if the individual is a security guard who makes regular checks on designated work locations or who must report the operating status of security equipment at specified intervals, the employee's routine fulfillment of these job duties may be sufficient evidence that he or she needs no further training in these duties, at least until new equipment, new technologies, or new procedures necessitate new training.

§ 37.43, "General Security Program Requirements" (continued)

§ 37.43(c), "Training" (continued)

§ 37.43(c)(3)
Refresher training must be provided at a frequency not to exceed 12 months and when significant changes have been made to the security program. This training must include:

§ 37.43(c)(3)(i)
Review of the training requirements of paragraph (c) of this section and any changes made to the security program since the last training;

§ 37.43(c)(3)(ii)
Reports on any relevant security issues, problems, and lessons learned;

§ 37.43(c)(3)(iii)
Relevant results of NRC inspections; and

§ 37.43(c)(3)(iv)
Relevant results of the licensee's program review and testing and maintenance.

§ 37.43(c)(4)
The licensee shall maintain records of the initial and refresher training for 3 years from the date of the training. The training records must include dates of the training, topics covered, a list of licensee personnel in attendance, and related information.

Explanation

The licensee must provide refresher training at least every 12 months and when significant changes have been made to the security program. The subsection specifies what the refresher training must include and states that licensees must keep training records for 3 years.

Q&As

Q1: How frequently should licensees provide refresher training?

A1: The regulation in 10 CFR 37.43(c)(3) requires the licensee to conduct refresher training, at a minimum, every 12 months. Licensees must also offer refresher training if the security program has changed significantly.

Q2: What should be included in refresher training?

A2: Developers of refresher training should review the training requirements of 10 CFR 37.43(c)(1) and 10 CFR 37.43(c)(2) and should address any changes to the security program since the last training, any changes in the assigned responsibilities of individual trainees that would require new training, recent information on any relevant security issues or lessons learned, relevant results from the program review or any NRC inspections, and relevant operating experience from the maintenance and testing program for security systems or system components. In addition, a licensee should consider providing refresher training after a security-related event so that affected or potentially affected employees can obtain a timely understanding of what happened and how to avoid or mitigate the consequences of a recurrence.

Q3: Under 10 CFR Parts 37.43(c)(3)(ii), (iii), and (iv), what things do the NRC consider "relevant" for purposes of refresher training?

A3: The term "relevant" is a common term and, in this case, simply refers to items that are related to security. Among examples of items refresher training should include are areas in which the staff has had trouble following the security requirements, areas in which corrective actions have not been effective or lessons learned have not addressed root causes of past or potential security deficiencies and new actions are required, violations of 10 CFR Part 37 security requirements or conditions adverse or potentially adverse to security that have been discussed in an inspection report, and measures necessary to fix any security issues that have been identified in the licensee's program review or testing and maintenance results.

Q4: When we give LLEA representatives a copy of our security plan, do we have to test them on their understanding?

A4: No. A licensee may, at its discretion, provide a copy of its security plan to the affected LLEA for information. However, the licensee is not required to provide a copy, nor does the NRC expect it to train an LLEA on its security plan. The plan itself applies only to the licensee and its employees and contractors.

§ 37.43, "General Security Program Requirements" (continued)

§ 37.43(d), "Protection of Information"

§ 37.43(d)(1)

Except as provided in paragraph (d)(9) of this section, licensees authorized to possess category 1 or category 2 quantities of radioactive material shall limit access to, and unauthorized disclosure of, their security plan, implementing procedures, and the list of individuals that have been approved for unescorted access.

§ 37.43(d)(2)

Efforts to limit access shall include the development, implementation, and maintenance of written policies and procedures for controlling access to, and for proper handling and protection against unauthorized disclosure of, the security plan and implementing procedures.

§ 37.43(d)(3)

Before granting an individual access to the security plan or implementing procedures, licensees shall:

§ 37.43(d)(3)(i)

Evaluate an individual's need to know the security plan or implementing procedures; and

§ 37.43(d)(3)(ii)

If the individual has not been authorized for unescorted access to category 1 or category 2 quantities of radioactive material, Safeguards Information, or Safeguards Information-modified handling, the licensee must complete a background investigation to determine the individual's trustworthiness and reliability. A trustworthiness and reliability determination shall be conducted by the reviewing official and shall include the background investigation elements contained in 10 CFR 37.25(a)(2) through (a)(7).

Explanation

Licensees must limit access to, and unauthorized disclosure of, their security plans, implementing procedures, and lists of individuals approved for unescorted access. Licensees must have written policies and procedures for the control of access to those documents and for the proper handling of, and protection against, unauthorized disclosure.

Licensees must evaluate an individual's need to know before allowing him or her access to the security documents. If the individual has not been authorized for unescorted access, he or she must undergo a background investigation to determine his or her T&R before being provided the protected information.

Q&As

Q1: What kinds of information are licensees required to protect under this subsection?

A1: Licensees are required to limit access to their security plans and implementing procedures to prevent unauthorized disclosure. Licensees also are required to limit access to their list of individuals approved for unescorted access. This includes the information contained in the protected documents.

Q2: Should I protect any forms of information other than hardcopies of sensitive information?

A2: Yes. The licensee should protect against any form of unauthorized disclosure of the protected information, including spoken and electronic disclosures.

Q3: Who is allowed access to this protected information?

A3: Licensees may only allow access to these documents to those individuals who have a need to know the information to perform their duties and who have been determined to be trustworthy and reliable based on the background investigation requirements that appear in 10 CFR 37.25(a)(2) through 10 CFR 37.25(a)(7). Licensee employees, agents or contractors, and employees of an organization affiliated with the licensee's company (e.g., a parent company) may be considered employees of the licensee for access purposes.

If a licensee sees any indication that the recipient would be unwilling or unable to provide proper protection for its sensitive information, it should not authorize that individual to receive it. Licensees must also ensure that individuals who are not authorized to receive such information do not overhear conversations related to the substantive portions of the sensitive information.

Q4: If I have a documented basis for having found an individual to be trustworthy and reliable for unescorted access to radioactive material under 10 CFR 37.25(a), can that individual have access to my security plan and procedures without redundant additional screening?

A4: Yes. However, the individual must have a need to know the information.

Q5: If I need to hire a new employee that will need access to the protected security information, can I accept a current background investigation from another licensee for a prospective new hire transferring from that licensee?

A5: Yes. Licensees are encouraged to share information about the current security authorization of employees who are departing to work for another licensee not only for access to radioactive material but also for access to sensitive information. This sharing of information can spare all licensees from having to conduct unnecessary additional background investigations. However, the regulations in 10 CFR Part 37 do not require the sharing of access authorization information. Such background information can only be shared on a voluntary basis by mutual consent. The licensee will still need to document the basis for the determination. Assuming that the existing background investigation is not older than 10 years, the licensee will not need to conduct a new reinvestigation.

Q6: What measures are required for adequate information protection?

A6: Licensees must develop, maintain, and implement written policies and procedures to ensure that only trustworthy and reliable individuals with a need to know are allowed access to security plans and procedures. See 10 CFR 37.43(d)(2). These policies and procedures must ensure the proper handling and protection of security plans and implementing procedures to protect against unauthorized disclosure. The licensee's policies and procedures should do the following:

- Include a general performance requirement that each person who produces, receives, or acquires the licensee's sensitive information ensures that such information is protected against unauthorized disclosure.
- Address how to protect sensitive information while in use, storage, and transit.

- Address the preparation, identification or marking, and transmission of documents or correspondence containing the licensee's security program information.
- Address how access to the licensee's security program information is to be controlled.
- Include methods for the destruction of documents that contain security program information.
- Include procedures for the use of automatic data processing systems that contain security program information.
- Address the removal of documents from the licensee's protected information category when they become obsolete or no longer sensitive.

§ 37.43, "General Security Program Requirements" (continued)

§ 37.43(d), "Protection of Information" (continued)

§ 37.43(d)(4)

Licensees need not subject the following individuals to the background investigation elements for protection of information:

§ 37.43(d)(4)(i)

The categories of individuals listed in § 37.29(a)(1) through (13); or § 37.43(d)(4)(ii)

Security service provider employees, provided written verification that the employee has been determined to be trustworthy and reliable by the required background investigation in § 37.25(a)(2) through (a)(7) has been provided by the security service provider.

§ 37.43(d)(5)

The licensee shall document the basis for concluding that an individual is trustworthy and reliable and should be granted access to the security plan or implementing procedures.

§ 37.43(d)(6)

Licensees shall maintain a list of persons currently approved for access to the security plan or implementing procedures. When a licensee determines that a person no longer needs access to the security plan or implementing procedures or no longer meets the access authorization requirements for access to the information, the licensee shall remove the person from the approved list as soon as possible, but no later than 7 working days, and take prompt measures to ensure that the individual is unable to obtain the security plan or implementing procedures.

§ 37.43(d)(7)

When not in use, the licensee shall store its security plan and implementing procedures in a manner to prevent unauthorized access. Information stored in nonremovable electronic form must be password protected.

§ 37.43(d)(8)

The licensee shall retain as a record for 3 years after the document is no longer needed: § 37.43(d)(8)(i)

A copy of the information protection procedures; and

§ 37.43(d)(8)(ii)

The list of individuals approved for access to the security plan or implementing procedures.

§ 37.43(d)(9)

Licensees that possess safeguards information or safeguards information-modified handling are subject to the requirements of § 73.21 of this chapter and shall protect any safeguards information or safeguards information-modified handling in accordance with the requirements of that section.

Explanation

This section lists the classes of individuals who are exempt from licensee background investigations for the protection of security information. The licensee must store security plans and implementing procedures to prevent unauthorized access and must password-protect information stored in nonremovable electronic form.

Information protection procedures and the list of individuals approved for access to the security plan or implementing procedures must be kept as records for 3 years. Licensees that possess SGI or SGI-M handling must protect it in accordance with 10 CFR 73.21.

Q&As

Q1: Is anyone relieved from the background investigation elements?

A1: Yes. The same categories of individuals who have been relieved from the background investigation elements for unescorted access to the radioactive material are also relieved from the background investigation elements for access to security information. See also the Q&As on 10 CFR 37.29. Under 10 CFR 37.29(a), these categories include the following:

- an employee of the Commission or the Executive Branch of the U.S. Government who has undergone fingerprinting for a prior U.S. Government criminal history records check
- members of Congress
- an employee of a member of Congress or Congressional committee who has undergone a prior U.S. Government criminal history records check
- the governor of a State or his or her designated State employee representative
- Federal, State, or local law enforcement personnel

- State radiation control program directors and State homeland security advisors or their designated State employee representatives
- Agreement State employees who conduct security inspections on behalf of the NRC under an agreement executed under Section 274.i of the AEA
- representatives of the IAEA engaged in activities associated with the U.S./IAEA Safeguards Agreement who have been certified by the NRC
- emergency response personnel who are responding to an emergency
- commercial vehicle drivers for road shipments of category 2 quantities of radioactive material
- package handlers at transportation facilities, such as freight terminals and railroad yards
- any individual who has an active Federal security clearance
- any individual employed by a service provider licensee for which the service provider licensee has conducted the background investigation for the individual and has approved the individual for unescorted access to category 1 or category 2quantities of radioactive material

Q2: How should a licensee protect sensitive information while in use, storage, and transit?

A2: The licensee should store the information in a locked cabinet, desk, or office. The licensee must password-protect information stored in nonremovable electronic form. Licensees must address how employees should protect the sensitive information while this information is in their possession, both at and away from work. Access to the keys, combinations, passwords, or other means used to secure the information needs to be limited to those persons authorized access to the information. Additional information on the protection of sensitive security-related information appears in NRC Regulatory Issue Summary 2005-31, "Control of Security-Related Sensitive Unclassified Nonsafeguards Information Handled by Individuals, Firms, and Entities Subject to NRC Regulation of the Use of Source, Byproduct, and Special Nuclear Material," dated December 22, 2005 (http://www.nrc.gov/reading-

Q3: How should a licensee prepare, identify, mark, and transmit documents or correspondence containing the licensee's security program information?

A3: A good practice is to prepare and mark licensee-generated security program information in such a manner as to ensure easy identification and proper handling. The front and back of folders that contain such information also should be marked for easy identification and proper handling.

Documents that do not, in themselves, contain security program information but are used to transmit one or more documents containing this information should be marked to indicate that security program information appears in the documents transmitted. Although security program information is not typically transmitted to the NRC, if a need for such transmittals exists, the licensee should mark them as follows: "Withhold from Public Disclosure in Accordance with 10 CFR 2.390." The licensee should place these markings at the top and bottom of only the first page of the transmitted document.

Q4: What methods should a licensee use for destruction of documents containing security program information?

A4: A good practice is to destroy documents that contain this information using a method that will prevent reconstruction of the information. Documents may be destroyed by tearing them into small pieces or by burning, pulping, pulverizing, shredding, or chemical decomposition. Similar methods may be used for film and microform media. ("Microform" is the generic term for any material on film or paper, such as a microfiche that contains a smaller reproduction of a document for transmission, storage, reading, and printing.) Electronic media, such as tapes, floppy discs, compact discs, digital video discs, and thumb drives, may be destroyed by shredding, smashing, burning, or chemical decomposition, and data on hard drives and servers should be erased, or the drives reformatted, before submitting them for recycling. (Note that regardless of the storage medium, the licensee should not send security program information to recycling without it being destroyed first.)

Q5: What methods should a licensee use to protect security information when using a data-processing system to store security program information?

A5: Individuals should protect the information during use by maintaining control and by ensuring that only individuals who have the "need to know" and who have been determined to be trustworthy and reliable have access to the information. One method that the licensee may use is to password-protect documents. The licensee also may use encryption techniques or place the documents on a restricted drive. Whatever method it uses, the licensee should describe the method in its information protection procedures.

Q6: When should a licensee remove documents from the licensee's protected information category if they become obsolete or no longer sensitive?

A6: A licensee may remove documents from its protected information category when it no longer needs to have a security plan and procedures (i.e., when it no longer needs to possess an aggregated category 2 or greater quantity of radioactive materials). Until that time, all the information in the licensee's security plan or procedures is unlikely to become obsolete or no longer sensitive. Neither the plan nor the procedures may be relieved from 10 CFR Part 37 information protection requirements if they retain any sensitive information, and even the original versions of the plan and procedures are likely to retain some sensitive information until the licensee no longer needs to protect a category 2 or greater quantity of material.

Q7: What documents do I need retain under this section? How long must I retain the documents?

A7: The licensee must retain a copy of the procedures on information protection. It also must keep a copy of the list of individuals approved to have access to the security information. These records must be kept for 3 years after the record is no longer needed or until the NRC terminates the license as required by 10 CFR 37.103.

Q8: Will licensees need to handle SGI?

AS: Some licensees will be required to handle SGI-M, which is a subset of SGI. The licensees considered to be handling this kind of security information include panoramic and underwater irradiators that possess greater than 370 TBq (10,000 curies) of byproduct material in the form of sealed sources, manufacturers and distributors of items that contain source material or byproduct or special nuclear material in greater than or equal to category 2 quantities of concern, and licensees that transport nuclear material in greater than or equal to category 1 quantities of concern. The security information that licensees develop and use is considered to be SGI-M and must be protected under the measures for SGI-M in 10 CFR 73.21 and 10 CFR 73.23.

Q9: Would an engineer designing the security systems for a blood irradiator room need to undergo a background investigation to have unescorted access to the irradiator room when it contains radioactive material?

A9: Yes. If the engineer is allowed unescorted access to the security zone containing the irradiator, he or she would be subject to the access authorization program in Subpart B of 10 CFR Part 37 and would need to undergo a background investigation. If the irradiator room was situated such that the engineer would not need to enter the security zone or will be escorted while within the security zone, he or she would not be subject to Subpart B.

If the engineer needed access to protected security information, 10 CFR 37.43(d)(4)(ii) permits a licensee to accept written verification from a security service provider that the provider has conducted the background investigation required in 10 CFR 37.25(a)(2) through 10 CFR 37.25(a)(7) and has determined its employee to be trustworthy and reliable. This would permit the security service provider's employee to have access to the protected information.

Q10: I have an employee who needs to know some components of my security plan to perform her job, but she doesn't need to have access to the plan itself or implementing procedures. Does she need to undergo a background check to have access to parts of my security plan?

A10: Yes. An insider with even a limited knowledge of the security plan or its implementing procedures may be able to use this knowledge or convey it to others for malevolent purposes. For example, a security guard with access only to an alarm-response schematic or an information technology specialist with access only to a communication system for security personnel could significantly compromise the security of radioactive material against theft, sabotage, or diversion. In addition, if an employee has become familiar with the plan and procedures, the licensee may need to conduct a background investigation and make a T&R determination even if the employee does not have actual access to the document. An individual already approved for access to radioactive material would not need a reinvestigation; however, he or she would need to have some knowledge of the security plan or procedures to fulfill a job responsibility to be authorized access to any part of these documents.

§ 37.45, "LLEA Coordination and Notification"

§ 37.45(a)

A licensee subject to this subpart shall coordinate, to the extent practicable, with an LLEA for responding to threats to the licensee's facility, including any necessary armed response. The information provided to the LLEA must include:

§ 37.45(a)(1)

A description of the facilities and the category 1 and category 2 quantities of radioactive materials along with a description of the licensee's security measures that have been implemented to comply with this subpart; and

§ 37.45(a)(2)

A notification that the licensee will request a timely armed response by the LLEA to any actual or attempted theft, sabotage, or diversion of category 1 or category 2 quantities of material.

Explanation

Licensees must coordinate, to the extent practicable, with an LLEA for responding to threats to the licensee's facility, including any necessary armed response. The licensee must provide a description of its facilities, radioactive materials, and security measures. The licensee must also state that it will request a timely armed response by the LLEA to any actual or attempted theft, sabotage, or diversion.

Q&As

Q1: What kinds of LLEA coordination activities are required?

A1: The coordination activities could involve meetings, telephone conferences, plant tours, training in radiation protection, tabletop exercises, and other communications to provide information. As discussed in more detail in the subsequent Q&As for this section, the licensee must notify the LLEA that it will request an armed response to any actual or attempted theft, sabotage, or malevolent diversion of radioactive material. The licensee must also provide information about the facility, its radioactive materials, and the licensee's current security measures. To facilitate a timely, coordinated, and effective response to a security incident, the licensee should verify site and LLEA contact information.

Q2: What is the purpose of the LLEA coordination?

A2: A timely and effective response to a security incident is part of an effective physical protection system. Because certain situations may necessitate an armed response, a strategy that is consistent in scope and timing with realistic potential vulnerabilities of the subject radioactive material should be coordinated well in advance with the LLEA. Another purpose of coordination is to provide the responsible LLEA with an understanding of the potential consequences associated with malevolent use of the radioactive material so that the LLEA can determine the appropriate priority of its response. The LLEA response would be necessary not only to interdict and disrupt an attempted theft or sabotage of the radioactive material but also for possible offsite coordination to protect public health and safety and to mitigate the potential consequences of a theft, sabotage, or malevolent diversion of radioactive material.

Q3: There are many different kinds of agencies responsible for providing security, including private security guard forces on the campuses of universities, hospitals, and other institutions. What kinds of law enforcement organizations should a licensee seek to coordinate with?

A3: The LLEA will need to be authorized by a Government entity to make arrests and to have the capability to provide an armed response. However, the LLEA does not need to be a municipal or a county police force. If a city or county hospital or State university campus police force is the nearest law enforcement agency to the licensee's operation capable of providing an armed response and making arrests, that police force would meet the definition of an LLEA.

A licensee also will need to consider if the LLEA could provide the needed armed response at all times, day or night, 7 days a week. Some LLEAs are on duty only during specified hours; in such cases, the licensee will have to identify and coordinate with the closest LLEA able to provide an armed response and arrest perpetrators when the primary LLEA is off duty.

Q4: Can an onsite proprietary professional security force with trained and armed officers be considered an LLEA?

A4: No. A private security force cannot be considered an LLEA unless it meets the criteria for an LLEA as defined in 10 CFR 37.5. An onsite armed force may provide the initial licensee response, but the licensee must still coordinate with an offsite LLEA unless the onsite force is also authorized by a Government agency to bear arms and make arrests, as many university campus police forces are. The offsite LLEA should still be notified of incidents immediately to provide additional assistance if necessary and to enable the LLEA to assess the potential for offsite impacts and the need to notify other agencies.

Q5: If I have a written agreement with a third party service that provides off-duty local law enforcement agents on site at all times, would that be acceptable to demonstrate compliance with the LLEA coordination requirement?

A5: Even if the individual off-duty police officers working for the third-party security service are authorized to carry firearms, to make arrests, and to provide an armed response to a security incident at a licensee's facility, the company itself would not qualify as an LLEA because it must be approved *as an agency* by a Federal, State, or local government to carry out these duties. Thus, an agreement with such a third party to provide security for the licensee's site is not, in itself, sufficient to demonstrate compliance with the coordination requirement. The licensee would still need to coordinate with an LLEA to request its assistance for either onsite or offsite support or both in the event of an attempted or actual theft, sabotage, or diversion. For example, the licensee would, among other activities, still need to coordinate with the LLEA under 10 CFR 37.45(d) at least every 12 months or when changes to its facility design or operation could adversely affect the potential vulnerability of its material to theft, sabotage, or diversion. The licensee may also want to consider whether and how the service provider's off-duty police officers could coordinate with their respective offsite LLEA employers in the event of a security incident at the licensee's site.

Q6: What kinds of information should a licensee provide to an LLEA for effective coordination?

A6: The information should include the important aspects of a licensee's physical protection program and other factors that would aid the LLEA to appropriately prioritize and respond to an alarm or other request by the licensee for response. As a good practice, examples of information that a licensee could discuss with the LLEA include, but are not limited to, the following:

- types and quantities of devices and radioactive material possessed
- potential hazards associated with loss of control of the devices and radioactive material
- specific facility information (i.e., contact information, floor plans, entrances, points of egress, or other information)
- site-specific physical protection measures that the licensee employs to delay an adversary from gaining access to the radioactive material
- established protocol for contacting the LLEA in response to an event
- licensee and LLEA points of contact for plans to recover stolen material that has been removed to an offsite location

At a minimum, the regulation in 10 CFR 37.45(a) requires information for coordination activities to include a description of the facility; the quantity and type of radioactive materials; the security measures in place at the licensee's facilities; and a notification that the licensee will request a timely and armed response to any actual or attempted theft, sabotage, or diversion of the licensee's radioactive materials. The regulation in 10 CFR 37.45(c) requires the licensee to document its coordination efforts, including the dates, times, and locations of meetings and a list of licensee and LLEA staff members present at the meetings.

Q7: Does the NRC require a licensee's security staff to have firearms?

A7: The rule does not require a licensee's security staff to be armed, nor does it prohibit licensees from arming their security staff. The licensee must comply with applicable State and local law.

Q8: What does the NRC mean by coordinating with an LLEA "to the extent practicable"?

A8: The NRC recognizes that some LLEAs may be reluctant to engage in coordination activities with a licensee, and the agency added "to the extent practicable" to avoid putting licensees into noncompliance for actions that they cannot reasonably be expected to control. This phrase allows a licensee to remain in compliance with the rule when a LLEA will not participate in any coordination activities. The licensee can comply with this requirement simply by documenting, in accordance with 10 CFR 37.45(c), that it has initiated and has periodically contacted the LLEA during the required timeframe, has provided information about its security needs and its intent to request an armed response, and has requested to coordinate a response to a security incident involving a category 2 or greater quantity of radioactive materials.

Q9: What should a licensee request as an adequate LLEA response to a threat to a licensee's facility? For example, would a single officer with radio backup be sufficient, or is an entire SWAT team necessary?

A9: The LLEA will decide what is adequate and will respond as appropriate to the event based on the agency's understanding of the situation and its potential consequences. One of the purposes of establishing liaison with the LLEA is to provide it an understanding of the potential consequences associated with the malevolent use of the radioactive materials subject to this rule so that the LLEA can appropriately determine the priority of its response. The LLEA response is necessary for offsite coordination to protect the public health and safety and to mitigate the potential consequences of the malevolent use of radioactive material.

Q10: Can LLEAs have access to the licensee's physical protection information?

A10: The licensee is *required* to give an affected LLEA some physical protection information. The regulation in 10 CFR 37.45(a)(1) requires the licensee to provide "a description of its facilities and the category 1 and category 2 quantities of radioactive materials along with a description of its security measures that have been implemented to comply with this subpart."

Q11: What are the LLEA's responsibilities for protecting this sensitive information?

A11: LLEAs are not subject to NRC information protection requirements; however, they may be subject to other information protection requirements under State or local law. Under 10 CFR 37.29(a)(5), State, local, and other law enforcement authorities are members of occupational groups deemed to be trustworthy and reliable by virtue of their employment status, and these authorities protect sensitive law enforcement information routinely in the course of their operations.

Q12: Would it be appropriate for a licensee to give a diagram of its facility to an LLEA?

A12: Depending on the size of the licensee's facility and the location of the at-risk material, providing a facility plan to the LLEA may be appropriate. The purpose of coordination is to provide the LLEA with the information that it deems necessary to do its job in responding to potential malevolent acts that involve lost, stolen, or missing radioactive material from the licensee's facility.

Q13: If a licensee is required to have an emergency plan under 10 CFR 30.32(i), can the licensee satisfy any of the requirements of 10 CFR 37.45 for LLEA coordination by complying with the emergency planning requirements of 10 CFR 30.32(i)(3)(xii)?

A13: Yes. The licensee can comply with the emergency planning requirements in 10 CFR 30.32(i)(3)(xii) to satisfy the requirements in 10 CFR 37.45 for LLEA coordination as long as its emergency plan specifically calls for LLEA coordination. The regulation in 10 CFR 30.32(i)(3)(xii) states, among other things, that an emergency plan under this section must include "provisions for conducting quarterly communications checks with offsite response organizations...to test response to simulated emergencies." The quarterly communications checks must include "the check and update of all necessary telephone numbers." Thus, if a licensee possesses at least a category 2 quantity of radioactive material and has an emergency plan that complies with 10 CFR 30.32(i) and if the plan specifies the nearest LLEA or LLEAs as "offsite response organizations" with which the licensee conducts quarterly communications checks, the licensee may cite the plan as one component of its demonstration of compliance with the general requirement under 10 CFR 37.45(a) to "coordinate, to the extent practicable, with an LLEA for responding to threats to the licensee's facility." These quarterly checks could also enable the licensee to demonstrate compliance with the requirement in 10 CFR 37.45(a)(2) for "[a] notification that the licensee will request a timely armed response by the LLEA to any actual or attempted theft, sabotage, or diversion," and compliance with the requirement in 10 CFR 37.45(d) to "coordinate with the LLEA at least every 12 months." Depending on the kinds of information in the emergency plan, it may also enable the licensee to comply with the requirement in 10 CFR 37.45(a)(1) to provide to the LLEA "a description of the facilities and the category 1 and category 2 quantities of radioactive materials along with a description of the licensee's security measures that have been implemented to comply with this subpart." However, the plan itself would not enable the affected licensee to comply with the notification requirements in 10 CFR 37.45(b), the documentation requirements in 10 CFR 37.45(c), or the requirement in § 37.45(d) to coordinate with the LLEA "when changes to the facility design or operation adversely affect the potential vulnerability of the licensee's material."

Q14: Is there some method in place whereby an LLEA may become informed about radioactive materials and the licensee's possession of such materials, including those in devices?

A14: Yes. The Federal Government through the U.S. Department of Homeland Security (DHS) is actively working with LLEA organizations to improve their awareness and response capabilities across a wide variety of threats by providing information, training, and funds. LLEAs can obtain additional information about training from the National Domestic Preparedness Consortium at http://www.ndpc.us/index.html, the local FBI Field Office Weapons of Mass Destruction Coordinator at http://www.fbi.gov/contact-us/field, and the

DHS National Network of Fusion Centers at http://www.dhs.gov/files/ programs /gc1156877184684.shtm.

In coordination efforts with the LLEA, the licensee's information should add to, and should help reinforce, the LLEA's knowledge of radioactive materials and the potential risks associated with both their legitimate and malevolent uses. Licensees can invite the LLEA to attend their annual health and safety and security training sessions.

Q15: Do I have to conduct response exercises with my LLEA or otherwise train my LLEA?

A15: No. The licensee is not required to offer training to the LLEA. Exercising the response portion of the licensee's security plan is a good practice, but it is not required. Some LLEAs may not be willing or able to participate, and the NRC also recognizes that requiring all licensees to exercise their response plans may not be cost effective for small licensees with less complex security plans.

Q16: What are the LLEA coordination requirements for work at a temporary jobsite?

A16: The rule does not require a licensee to coordinate with an LLEA for work at temporary jobsites. However, licensees are not precluded from coordinating with the LLEA as an extra margin of security for work at temporary jobsites. For licensees that routinely work at temporary jobsites for months at fixed locations, the licensee may want to identify and notify the closest LLEA for the benefit of both parties even though it is not required.

§ 37.45, "LLEA Coordination and Notification" (continued)

§ 37.45(b)

The licensee shall notify the appropriate NRC regional office listed in § 30.6(a)(2) of this chapter within 3 business days if:

§ 37.45(b)(1)

The LLEA has not responded to the request for coordination within 60 days of the coordination request; or

§ 37.45(b)(2)

The LLEA notifies the licensee that the LLEA does not plan to participate in coordination activities.

Explanation

The licensee must notify the appropriate NRC regional office within 3 business days if the LLEA has not responded to its request for coordination within 60 days or if the LLEA notifies the licensee that it does not plan to participate in coordination activities.

Q&As

Q1: Would an LLEA's decision not to coordinate put a licensee into noncompliance with these coordination requirements?

A1: No. The NRC recognizes that it cannot exercise authority over LLEAs or any party over which a licensee has no control and over which the NRC has no legal jurisdiction. The NRC also recognizes that an LLEA may have good reasons, including resource limitations and the possibility of other coinciding incidents within its jurisdiction, for not engaging in coordination activities with a licensee.

Q2: What happens when an LLEA declines to coordinate with a licensee?

A2: A licensee must notify the NRC whenever an LLEA with jurisdiction over the licensee's facilities declines to engage in coordination activities.

Q3: To trigger the requirement to notify the NRC, must a licensee have a written or oral statement from an LLEA stating that it does not plan to participate in coordination activities?

A3: No. Because the rule requires the licensee to notify the NRC if the LLEA has not responded within 60 days to a licensee's request for coordination, a licensee does not need to have an official statement from the LLEA that it does not plan to participate. The LLEA's actions can demonstrate that the LLEA does not plan to coordinate with the licensee.

Q4: How should a licensee notify the NRC regional office to meet 10 CFR 37.45(b) requirements?

A4: The licensee may notify the NRC regional office by letter, facsimile, e-mail, or telephone.

The regulation in 10 CFR 30.6 provides contact information. If the licensee contacts the NRC by telephone, it should document the conversation so that it will have a record of the notification.

Q5: Some State or local jurisdictions may have a requirement that the initial response to any emergency involving radioactive materials must be provided by other than armed LLEA personnel, such as hazmat responders. How do I know whether such a requirement affects my LLEA's response to a security incident?

A5: The licensee should ask the LLEA. A convenient time to ask could be during one of the licensee's required interactions with the LLEA to develop or maintain effective coordination of responses to a security incident. If such a State or local requirement exists, the licensee should notify the designated first responder organization about the nature of its emergency response needs during a security incident so that unarmed and unsuspecting responders are not needlessly put in harm's way. The licensee may also ask the hazmat or other designated first responder to coordinate with the LLEA to develop an agreed upon response protocol that would allow the licensee to contact the LLEA first for a response to a security incident involving its radioactive materials.

§ 37.45, "LLEA Coordination and Notification" (continued)

§ 37.45(c)

The licensee shall document its efforts to coordinate with the LLEA. The documentation must be kept for 3 years.

Explanation

Self-explanatory.

Q&As

Q1: Besides the documentation of meetings and correspondence, what other information on LLEA coordination activities should a licensee document?

A1: A licensee should also document any written agreement or prearranged response plan with an LLEA if the LLEA is willing to enter into such an agreement or plan.

Q2: In cases where an LLEA has decided not to participate in coordination activities, how should a licensee document that it has made a good faith effort to coordinate with that LLEA? Must a licensee send all correspondence to such an LLEA by certified mail, UPS, or other third-party delivery method in order to produce a verifiable record?

A2: No. Using certified mail, UPS, or another third-party delivery method to document the licensee's transmittal of a coordination request would be an acceptable way to comply with the requirement; however, the licensee does not need to use such methods if it keeps a copy of the dated letter, e-mail, or other correspondence and maintains records of calls to the LLEA.

Q3: Must a licensee maintain paper records of its LLEA coordination activities and correspondence? What about e-mails and other electronic communications, including digitized images, and documents generated on a computer?

A3: No. The licensee does not need to maintain records in paper form. As long as the record is legible and retrievable, the licensee can keep a reproduced copy, microform copy, or electronic copy. The regulation in 10 CFR 37.45(c) requires licensees to maintain records of LLEA coordination activities for a period of 3 years.

§ 37.45, "LLEA Coordination and Notification" (continued)

§ 37.45(d)

The licensee shall coordinate with the LLEA at least every 12 months, or when changes to the facility design or operation adversely affect the potential vulnerability of the licensee's material to theft, sabotage, or diversion.

Explanation

The licensee must coordinate with the LLEA every 12 months.

Q&As

Q1: When should I coordinate with the LLEA concerning changes to my facility's design or operation?

A1: The licensee should notify the LLEA as soon as possible after making any changes to the facility design or operation that could affect the LLEA's response to an incident. If possible, the licensee should notify the LLEA before making such changes. In addition, a good practice is to afford LLEA response personnel the opportunity to familiarize themselves with the facility once any design or operating changes have been made.

Q2: I understand that I must coordinate with the LLEA at least every 12 months even if there are no significant changes to my facility's design or operation. What topics or issues should I cover when I contact the LLEA in one of these 12-month coordination activities?

A2: The coordination may include, but should not be limited to, a discussion of any recent security developments since the licensee's last coordination meeting, the potential consequences of any planned or potential changes in either its facilities or operations, or the LLEA's capabilities or policies. Thus, the licensee should be prepared to review any security events or trends over the preceding 12 months that could have involved the LLEA and to review any security events or trends for which the licensee has taken or plans to take corrective action. The licensee should also be prepared to discuss any changes in its facility's existing or planned security measures, policies, notification procedures, equipment, or other resources that could affect the LLEA's response or ability to respond to an actual or attempted theft, sabotage, or diversion of radioactive materials at the licensee's facilities over the upcoming 12 months. An example of such a change would be a revision of the licensee's security plan or procedures; a new monitoring, alarm, or communication system; or a decision to procure security services from a commercial contractor.

The licensee should also discuss any changes in its LLEA point-of-contact information either at its facility or at the LLEA. In addition, depending on the licensee's and the LLEA's available time and resources, the licensee could offer the LLEA a tour of its facility or an opportunity to participate in planned drills or tabletop exercises or both activities.

Q3: Must I convene a face-to-face meeting for 12-month coordinations?

A3: No. Although a face-to-face meeting with LLEA officials would be desirable, the NRC understands that competing demands on the LLEA's time may make such meetings difficult to arrange.

Q4: Who should participate in these coordination activities?

A4: At a minimum, participants on the licensee's side should include the individual who has operating responsibility for implementation of the security plan or his or her designee and the individual that the licensee identified to the LLEA as its point of contact for coordination activities, if the two responsibilities are not assigned to the same person. If the participants need a prompt decision from an individual at a higher level of licensee management, that individual should attend. Depending on current demands for LLEA assistance, getting the desired LLEA representative to attend may be difficult; however, the licensee should, for consistency, try to ensure that at least the designated LLEA point of contact, if there is one, participates.

§ 37.47, "Security Zones"

§ 37.47(a)

Licensees shall ensure that all aggregated category 1 and category 2 quantities of radioactive material are used or stored within licensee-established security zones. Security zones may be permanent or temporary.

Explanation

All aggregated category 1 and category 2 quantities of radioactive material must be used or stored within licensee-established permanent or temporary security zones.

Q&As

Q1: How do I determine whether I possess an aggregated category 1 or category 2 quantity of radioactive material?

A1: The sum-of-fractions method, also known as the unity rule, is used to determine if the licensee possesses a category 1 or category 2 quantity of radioactive material. A licensee may need to implement 10 CFR Part 37 requirements even if it does not possess any single source or single radionuclide in excess of a category 2 threshold. For examples of unity rule calculations, see Q1 and A1 under the definitions of category 1 and category 2 quantities in 10 CFR 37.5 of Subpart A.

Q2: What is a security zone, and what is its purpose?

A2: A security zone is an area, defined by the licensee, that both isolates and controls access to category 1 or category 2 quantities of radioactive material. The purpose of the zone is to define the area that contains the quantities of material that must be protected. A security zone effectively defines the areas in which the licensee will apply the isolation and access control measures required by 10 CFR Part 37. The licensee must use or store all category 1 and category 2 quantities of radioactive material only within a security zone.

Q3: Can I use an area established for radiation safety to meet the requirements for a security zone? Can they be the same?

A3: Depending on the licensee's safety and security needs, the boundaries of these two types of areas may differ or may be identical. Because the purpose of the security zones is different from the radiation safety purposes of the restricted areas and controlled areas defined in 10 CFR 20.1003, the security zone does not have to be the same as either of these areas; however, it may be the same.

However, because measures to control access are required for both radiation protection and security, a licensee does have the flexibility to use an area required for radiation protection purposes to fulfill the required functions of a security zone. The intent of 10 CFR 37.47 is to allow licensees flexibility in establishing access control and isolation in a manner that would allow reliance on existing systems and procedures, which are already in use for radiation protection, for physical security. For example, consider a temporary well-logging operation for which the regulations in 10 CFR 39.71, "Security," require the licensee to have a "restricted area" to "maintain direct surveillance...to prevent unauthorized entry." A licensee could define a security zone with the same boundaries as this "restricted area," which is defined in § 20.1003 as "an area, access to which is limited by the licensee for the purpose of protecting individuals against undue risks from exposure to radiation and radioactive materials." In this case, the restricted area required for radiation protection may also serve to provide security from exposure to this radiation.

Similarly, a radiographer could choose to define a security zone with the same boundaries as the "high radiation area" over which the regulations in 10 CFR 34.51, "Surveillance," require radiography licensees to "maintain direct visual surveillance...to protect against unauthorized entry." (As defined in 10 CFR 20.1003, a "high radiation area" is "an area, accessible to individuals, in which radiation levels from radiation sources external to the body could result in an individual receiving a dose equivalent in excess of a specified dose rate.")

Q4: How does the NRC define a "temporary" security zone?

A4: The NRC considers a temporary security zone to be one established by the licensee to provide physical protection for a category 1 or category 2 quantity of radioactive materials for a specified, limited expected time, which may be several minutes to several months, to accomplish a specified task.

Q5: How does a permanent security zone differ from a temporary one?

A5: A permanent security zone uses permanent barriers to provide isolation and to aid in controlling access to the material. These barriers may consist of fences, gates, free-standing walls for exterior areas, exterior or interior building walls, doors, locked windows, bars, or grillwork. The barriers or walls control access to the security zone through established access control points, deter and delay penetration by an unauthorized person or unauthorized people, and aid detection by providing an indication of forced penetration.

Temporary security zones do not need to have permanent barriers at their boundaries; however, licensees can use permanent barriers if they are available at the location of the work. The licensee may use other devices that warn passersby of the restricted nature of an area to provide isolation of the material and the people using the material in a temporary security zone. The licensee can also accomplish access control through surveillance of the material, persons, and area by an authorized individual.

Q6: If I'm an NRC licensee servicing a sealed source or other device in an Agreement State and my customer's Agreement State regulatory agency hasn't yet adopted 10 CFR Part 37 compatible implementing regulations, am I still required to follow the 10 CFR Part 37 requirements of my license?

A6: Yes. The licensee must follow the security requirements in its license and those of the Agreement State in which it works. Although the licensee will have to meet the requirements in its NRC license and its customer's Agreement State license, the points at which these requirements apply may differ according to if and when possession of the source changes hands. For example, if the licensee is doing maintenance that does not require it to remove the source and to take possession of it for offsite shipment, it will be subject to the applicable access control, monitoring, documentation, and other requirements in its customer's Agreement State license. However, if the licensee brings a replacement source onto its customer's site, the requirements of the licensee's license apply along with those of the customer's license, when the licensee arrives onsite, until it has installed the replacement source in the customer's device and has completed any remaining testing and maintenance required by its license and until the customer has taken possession of the source. In any case, both the licensee and the customer must have a clear understanding from the outset about which one (the licensee or the customer) is responsible for compliance with which requirement. For example, both entities should know which one is responsible for calling an LLEA in the event of a theft, sabotage, or diversion. However, if the licensee's Agreement State customer believes that compliance with the licensee's license would cause a violation of the customer's license requirements, the licensee and the customer should contact the regional NRC office and the appropriate Agreement State regulatory agency for further guidance.

§ 37.47, "Security Zones" (continued)

§ 37.47(b)

Temporary security zones must be established as necessary to meet the licensee's transitory or intermittent business activities, such as periods of maintenance, source delivery, and source replacement.

Explanation

Licensees must establish temporary security zones as necessary for time-limited or intermittent business activities, such as source maintenance, delivery, and replacement.

Q&As

Q1: When should I establish a temporary security zone?

A1: The licensee should establish a temporary security zone whenever it has a periodic or unplanned short-term operating requirement for the use or storage of a category 1 or category 2 quantity of radioactive material outside of a permanent security zone. Examples of such a requirement could include a need to repair, maintain, or calibrate a radioactive source or a device that contains a source at a location in which the licensee does not routinely use or store the source or device. The licensee should also establish a temporary security zone when it takes delivery of a source or device outside of its established permanent security zone. Additionally, work at a temporary jobsite will require the establishment of a temporary security zone at the jobsite.

Q2: How do I establish a temporary security zone?

A2: Although the licensee may be able to take advantage of permanent barriers if such barriers are available at the site, it may establish a security zone simply by keeping the area under direct supervision and control by an approved individual or individuals.

Q3: May I station someone who is not an approved individual outside a temporary security zone with a radio or other means of communication to notify people inside the zone about an impending intrusion?

A3: The rule does not prevent the licensee from using an unapproved individual to provide additional continuous direct surveillance; however, the regulation in 10 CFR 37.47(c)(2) does not allow the licensee to rely on any individual who has not been approved for unescorted access to be a monitor of access to a temporary or permanent security zone unless an approved individual monitors the unapproved individual. Allowing an individual who has not been determined trustworthy and reliable under this part to monitor a security zone without surveillance by an approved individual could provide an opportunity for insiders to compromise the security of the material.

Q4: How large should a temporary security zone be?

A4: The rule does not specify the size of a security zone. Locations are configured differently and do not lend themselves to generically defined physical areas; the security zone concept permits significant flexibility for licensees to account for a range of site-specific concerns. The licensee will determine its physical dimensions on an individual basis, and these dimensions can change based on operating status and security needs. The security zone may be as small as a locked cabinet or as large as a warehouse area.

Q5: If I routinely work in a large plant facility and move frequently from one location to another within the same facility, do I have to establish a different temporary security zone for each location, or can I consider the whole facility to be a temporary security zone that encompasses all my onsite work locations?

A5: The licensee's decision about the proper size of the temporary security zone is its alone to make, because only the licensee can know the conditions at the site and the requirements of its work. However, in making its decision, the licensee will need to consider the basic performance requirements for establishing a security zone. (See 10 CFR 37.47(c).) The regulation in 10 CFR 37.47(c) requires the licensee to isolate the subject radioactive materials to allow unescorted access only to approved individuals, using either direct control of the security zone by approved individuals at all times, continuous physical barriers that allow access to the security zone only through established access control points, or a combination of continuous physical barriers and direct control. For example, if the facility is small enough to provide continuous physical barriers that allow access only through an established access control point that the licensee controls, it may be able to consider the entire facility a temporary security zone. For those spaces in which the licensee will be using its source or device between access-controlled permanent zones, its temporary security zone does not need to be any larger than the space in which it needs to control access.

Q6: How do I meet 10 CFR Part 37 requirements for controlling unescorted access to a temporary security zone if I'm at a temporary jobsite that doesn't have such continuous physical barriers as lockable doors?

A6: When work needs to be done inside a temporary security zone that does not have physical barriers or special equipment, a licensee could meet the requirements for controlling unescorted access by having the material, persons, and area within the zone under the direct control of approved individuals at all times.

Q7: When would a temporary security zone have to be considered permanent?

A7: The rule does not specify a time limit by which a temporary security zone must become permanent; however, 10 CFR 37.47(b) does require a temporary security zone "necessary to meet the licensee's transitory or intermittent business activities, such as periods of maintenance, source delivery, and source replacement." Most licensees will probably have a business interest in maintaining temporary security zones for only a limited time because the cost of isolating radioactive material behind permanent physical barriers is likely to be cheaper than the labor cost of maintaining continuous direct surveillance.

§ 37.47, "Security Zones" (continued)

§ 37.47(c)
Security zones must, at a minimum, allow unescorted access only to approved individuals through:

§ 37.47(c)(1)
Isolation of category 1 and category 2 quantities of radioactive materials by the use of continuous physical barriers that allow access to the security zone only through established access control points. A physical barrier is a natural or manmade structure or formation sufficient for the isolation of the category 1 or category 2 quantities of radioactive material within a security zone; or

§ 37.47(c)(2)
Direct control of the security zone by approved individuals at all times; or

§ 37.47(c)(3)
A combination of continuous physical barriers and direct control.

Explanation

Security zones must allow unescorted access only to approved individuals by either isolating the radioactive materials with continuous physical barriers that allow access only through established control points, by controlling the security zone directly by approved individuals at all times, or by using a combination of physical barriers and direct control.

Q&As

Q1: For the purposes of 10 CFR Part 37, what is an "approved individual"?

A1: As defined in 10 CFR 37.5 of this part, an approved individual is an individual whom the licensee has determined to be trustworthy and reliable in accordance with Subpart B of this part and who has completed the training required by 10 CFR 37.43(c).

Q2: For the purposes of 10 CFR Part 37, what is a "continuous physical barrier"?

A2: A continuous barrier must limit access to the security zone only through one or more established access control points. The continuous barrier should have no openings other than access control points, including windows, large enough to allow a person to enter the security zone and bypass the access control point. For example, a wall should be continuous from the floor to the structural ceiling, and openings (such as vents) that are greater than 96 square inches, whereby the smallest dimension is greater than 6 inches, should have metal grates, bars, expanded metal (i.e., an industry term for a screen made of steel or similarly strong metal through which an observer can see activities inside or outside an enclosure), or some other barrier that cannot be removed from outside the security zone.

Q3: If I need to use locks, what controls are required to limit access to the keys to my facilities?

A3: The licensee must limit access to the keys to locks for security zones and category 1 or category 2 materials or devices within those zones solely to those individuals who must have access to perform their assigned tasks and who have been determined trustworthy and reliable in accordance with Subpart B. However, a better security practice is to avoid relying on keys to control access to the security zone. Key control becomes more difficult as time passes (especially for large institutions) as more duplicate keys are made for newly approved individuals. The same is true for the control of lock combinations over time. Key cards, cipher locks, or some other type of electronic access control device should be used where feasible instead of keyed or combination locks. Additionally, each approved individual should have a unique access code. With these devices, the licensee can delete the access code after an individual no longer requires access to the security zone or when a key card is lost or stolen.

Q4: Does 10 CFR Part 37 require that locks for different rooms be rekeyed to different keys or combinations?

A4: No. However, different keys or lock combinations for different rooms would improve security by making it more difficult for an adversary to aggregate radioactive material into a category 1 or category 2 quantity. The licensee must determine how it will control access to the facility.

Q5: Shouldn't the features of a device containing a category 1 or category 2 quantity of radioactive material be given credit for providing some isolation for purposes of access control?

A5: No. The NRC wrote the requirements for 10 CFR Part 37 with full awareness of the features of devices that licensees commonly use and the quantities of licensed material these devices typically contain. The devices are designed to limit access to the sources for radiation protection purposes, whereas the regulations in 10 CFR Part 37 also require the control of access to devices that contain the radioactive material. The requirements were designed to provide a defense-in-depth strategy for the protection of radioactive material in category 1 or category 2 quantities. No single control can provide the same level of protection as the combination of controls required by 10 CFR Part 37.

Q6: Self-contained irradiators have shown themselves to be very safe for day-to-day use without operators having access to the radioactive material. Controlling access to this material is generally much easier than controlling access to the irradiator itself. Are the 10 CFR Part 37 requirements expected to address access to the radioactive material or to the irradiator?

A6: Both. To address potential misuse with malevolent intent, the NRC designed the requirements for 10 CFR Part 37 to control access to both the radioactive material and the irradiator by controlling access to the security zone. The NRC has engaged the expertise of national laboratories that have shown that these devices may be vulnerable to theft, sabotage, or diversion under certain scenarios. For this reason and for the possibility that a trained insider could assist malevolent acts, the NRC has determined that certain additional security measures are necessary in the current threat environment. The regulations in 10 CFR Part 37 requires a layered, defense-in-depth approach to enhance the security of radioactive material in category 1 and category 2 quantities. No single measure can provide the required security for this material; therefore, a licensee must implement all applicable 10 CFR Part 37 requirements unless the licensee requests, and the NRC approves, an exemption.

Q7: Can the requirements for controlling access be waived if compliance is overly burdensome?

A7: A licensee may seek, and the NRC may grant, an exemption from any NRC requirement on a case-by-case basis; however, the NRC is unlikely to grant relief from an access control requirement simply because the licensee considers it burdensome.

Q8: What if a licensee's truck carrying a category 1 or category 2 quantity of radioactive material breaks down and must be towed to a shop for repairs, and the repair shop doesn't allow licensee personnel into the repair bay to keep the onboard security zone under continuous surveillance?

A8: The licensee must still meet the applicable requirements of 10 CFR Part 37 to ensure that the sources and their shielding devices are not removed. In this case, an approved individual could lock the truck, remain at the shop, and keep the truck in sight to control and maintain constant surveillance of licensed material. Alternatively, the licensee could remove the device and maintain control of the material until the truck is repaired or until the source(s) can be returned to a licensed facility.

Q9: Can the requirements of 10 CFR 37.53 for independent physical barriers for mobile sources be used to control access and prevent an unescorted individual from entering a room where a radioactive source is located?

A9: No. The intent of 10 CFR 37.53 is not to achieve total access control but to provide additional delay in removing a mobile device from the facility, temporary jobsite, or vehicle. For mobile devices, delay barriers are required *in addition to* the access controls required by 10 CFR 37.47. Without the additional barriers, a portable or mobile device could more easily be stolen, sabotaged, or diverted.

Q10: If I have a room with a conventional door lock and entry alarms, what else do I need for an effective security zone?

A10: To protect against unauthorized access to, and removal of, material behind conventional door locks with entry alarms, the licensee may need to consider additional means, such as guards, closed-circuit television, or motion detectors. The specific system and the means to protect material is left to the licensee's discretion; however, to meet 10 CFR Part 37 requirements, the licensee should consider reasonably foreseeable actions by adversaries and the methods that they could use to gain unauthorized access. For example, walls should be continuous from floor to ceiling. Vents and other openings greater than 96 square inches, whereby the smallest dimension is greater than 6 inches, should have metal grates, expanded metal, bars, or some other barrier that cannot be removed from outside the security zone. For additional technical guidance on the capabilities and applications of different access control system technologies, the licensee may refer to NUREG-1964, "Access Control Systems: Technical Information for NRC Licensees," issued April 2011.

Q11: For facilities that may store radioactive materials or devices, such as dosimetry calibrators, outside a protected area, what constitutes an adequate physical barrier? Would an unlocked concrete container be adequate, for example, if it could not be opened without the use of a mobile crane?

A11: The adequacy of a barrier around a storage area would depend on site-specific factors that the licensee should consider. For example, these factors should include the configuration of the storage area and whether it abuts and uses the perimeter fence at the site boundary; the proximity of the radioactive materials to the fence and their vulnerability to theft or diversion; and the proximity of other means of access, such as vehicles, cranes, ladders, or other tools that could readily be used to defeat the barrier.

In assessing the adequacy of any physical barrier, the licensee should consider, as generally good practice, its contribution to an integrated security system designed to provide defense in depth through effective detection, assessment, and response to a security event without delay. The licensee is responsible for ensuring that any barrier will perform as designed to enable the security zone as a system to meet the performance objective of isolating the material and allowing access only through established access control points under 10 CFR 37.47(c)(1). The licensee may also want to consider if a barrier could delay intruders sufficiently to contribute to the overall security program's ability to meet its performance objective of detecting, assessing, and responding to an actual or attempted unauthorized access "without delay" under 10 CFR 37.41(b).

Q12: How do I meet 10 CFR Part 37 requirements for controlling unescorted access to a temporary security zone if I'm at a temporary jobsite that doesn't have physical barriers or special equipment, such as lockable doors?

A12: A licensee could meet the requirements for a temporary security zone simply by keeping the area under "direct supervision" by authorized personnel.

§ 37.47, "Security Zones" (continued)

§ 37.47(d)

For category 1 quantities of radioactive material during periods of maintenance, source receipt, preparation for shipment, installation, or source removal or exchange, the licensee shall, at a minimum, provide sufficient individuals approved for unescorted access to maintain continuous surveillance of sources in temporary security zones and in any security zone in which physical barriers or intrusion detection systems have been disabled to allow such activities.

Explanation

Licensees must provide sufficient individuals approved for unescorted access to maintain continuous surveillance of category 1 quantity sources when physical barriers or intrusion detection systems (IDSs) have been disabled.

Q&As

Q1: When are special additional measures for category 1 quantities of radioactive material required?

A1: During equipment maintenance or source receipt, installation, replacement, or preparation for shipping, tamper-indicating devices and other intrusion detection equipment typically must be disabled to permit the device to be opened without tripping alarms. After replacement, the removed sources must be prepared onsite for shipment back to the manufacturer or for storage and eventual disposal. These nonroutine operations during a time when devices for detecting theft or diversion are disabled require additional measures to compensate for the temporary increase in vulnerability.

Q2: Does the approved individual who must perform the continuous surveillance of the source replacement or maintenance work have to remain on duty for the duration of the work?

A2: No. An approved individual must perform the continuous surveillance; however, he or she does not need to be the same individual for the duration of the work. If the original approved individual who is observing the work must be relieved, the second observer must begin surveillance before the original observer departs.

> **§ 37.47, "Security Zones" (continued)**
>
> **§ 37.47(e)**
> Individuals not approved for unescorted access to category 1 or category 2 quantities of radioactive material must be escorted by an approved individual when in a security zone.

Explanation

An approved individual must escort an individual who has not approved for unescorted access when he or she is in a security zone.

Q&As

Q1: What is an escort's responsibility?

A1: The objective of escorting is to maintain effective control of access to protected radioactive material, including access by unapproved individuals who require access to a security zone to perform their duties. The escort must maintain continuous surveillance of the escorted individual(s) and must carry out assigned responsibilities for detection, assessment, and response as required. The purpose of the escort is to ensure that escorted individuals perform their duties as intended and do not misuse their access to the protected radioactive materials. Procedures for escorts should be tailored to each facility's operations. Licensees should consider limiting the number of people assigned to an escort to avoid degrading the escort's effectiveness in completing other assigned safety and security responsibilities.

Q2: Do licensees have to visually distinguish (e.g., with badges) all individuals who have not been granted unescorted access?

A2: No. This is not a requirement; however, licensees should consider methods for distinguishing individuals approved for unescorted access from those requiring escort. For example, those individuals approved for unescorted access to category 2 or greater quantities of radioactive material could wear colored badges or other identifying articles. Colored badges or identification cards may be appropriate for a larger organization, whereas simple face recognition may be appropriate in a smaller one. The ability to visually distinguish approved individuals from those who require escort is considered a best security practice, although other ways to distinguish these individuals may be available, such as by electronically coded badges. In any case, the method used to distinguish individuals who require escort should help facility personnel in the early detection and timely assessment of unauthorized access incidents.

Q3: Can we assume that patients can be granted unescorted access during treatments?

A3: No. Patients who are being treated with devices containing category 1 or category 2 quantities of radioactive material must be escorted. The escort does not need to be physically with the patient in the treatment room; however, he or she can observe from outside the room. The licensee should restrict patients' access to loaded teletherapy devices during their treatment with these devices, and the licensee must comply with 10 CFR Part 37 requirements.

The licensee should not permit patients access to areas in which radioactive material is stored without being escorted by approved individuals.

Q4: Must an escort always maintain line-of-sight surveillance of an unapproved individual in a security zone?

A4: No. However, the escort must maintain "direct continuous visual surveillance," which he or she may conduct using video surveillance in some, but not all, cases. For example, video surveillance of patients during a treatment is appropriate; however, video surveillance in which the "escort" is in another building would not be appropriate.

Q5: Must the escort always be physically present when an individual who is not approved for unescorted access is in the security zone?

A5: Generally, yes. However, an escort may monitor an unapproved individual remotely (e.g., using a closed-circuit TV system or a window into the room) as long as the escort maintains continuous visual surveillance over the individual at all times. Under some circumstances (e.g., when a patient undergoes teletherapy), remote surveillance might be preferable to the actual physical accompaniment of an unapproved individual.

Q6: How can a single individual, such as a medical technologist assigned to be an escort, maintain "direct control" of the security zone sufficiently to prevent an incident or even raise an alarm in the event that someone is armed and intent upon gaining access to radioactive material?

A6: An absolutely assured method to prevent a determined, well-conceived, and well-equipped effort to gain unauthorized access to radioactive material does not exist. The purpose of the escort is to identify, assess, and respond if possible to the unauthorized activities by a visitor or another individual who has not been granted unescorted access. The escort should also observe behavior that may suggest an interest in defeating the security system. For example, the escort should notice if the visitor shows an unusual interest in the system, equipment, and procedures used to protect the security zone.

§ 37.49, "Monitoring, Detection, and Assessment"

§ 37.49(a), "Monitoring and Detection"

§ 37.49(a)(1)

Licensees shall establish and maintain the capability to continuously monitor and detect without delay all unauthorized entries into its security zones.

Licensees shall provide the means to maintain continuous monitoring and detection capability in the event of a loss of the primary power source, or provide for an alarm and response in the event of a loss of this capability to continuously monitor and detect unauthorized entries.

Explanation

Licensees must continuously monitor and detect without delay all unauthorized entries into security zones. The licensee must maintain a continuous monitoring and detection capability whenever the primary power source is lost, or the licensee must provide for an alarm and response when the capability to continuously monitor and detect unauthorized entries is lost.

Q&As

Q1: Is there a need to provide security monitoring for locations other than windows, doors, and access ways?

A1: Yes. The detection system must be able to detect all unauthorized access to the security zone, including breaches of barriers used to isolate and control access to the protected radioactive material. This capability may require security monitoring beyond that installed on windows, doors, and access ways.

Q2: If a licensee opts to provide an alarm system for a loss of the primary power source for monitoring and detection systems, must the system be calibrated to alarm for power surges, brownouts, or other anomalies in the electricity supply system? What would constitute a “loss” of the primary power source? What other sources of information can I use for monitoring and detection methods and equipment?

A2: The licensee should calibrate an alarm system for a power source impairment to trip for power surges, brownouts, and other anomalies that would cause a loss of the monitoring or detection system’s functionality. A licensee should consider a “loss” of its primary power source to be any anomaly that impairs the ability of a monitoring or detection system to perform as expected. If the licensee chooses instead to use an alternate or auxiliary power source, such as a gasoline-fueled generator, the alternate power source should be set to power the monitoring and detection system automatically.

Q3: What other sources of information can I use for monitoring and detection methods and equipment?

A3: For additional technical information on the capabilities and applications of different backup power intrusion detection technologies, the licensee may refer to NUREG-1959, “Intrusion Detection Systems and Subsystems: Technical Information for NRC Licensees,” issued March 2011. For technical information on the capabilities and applications of various access control systems, the licensee may refer to NUREG-1964.

§ 37.49, "Monitoring, Detection, and Assessment" (continued)

§ 37.49(a)(2)
Monitoring and detection must be performed by:

§ 37.49(a)(2)(i)
A monitored intrusion detection system that is linked to an onsite or offsite central monitoring facility; or

§ 37.49(a)(2)(ii)
Electronic devices for intrusion detection alarms that will alert nearby facility personnel; or

§ 37.49(a)(2)(iii)
A monitored video surveillance system; or

§ 37.49(a)(2)(iv)
Direct visual surveillance by approved individuals located within the security zone; or

§ 37.49(a)(2)(v).
Direct visual surveillance by a licensee designated individual located outside the security zone.

Explanation

The licensee must perform monitoring and detection by either a monitored IDS linked to a central monitoring facility, electronic intrusion detection alarms that will alert nearby facility personnel, a monitored video surveillance system, or direct visual surveillance.

Q&As

Q1: Would implementation of an area monitor connected to a silent alarm, key card access to the area, and a video monitoring system be adequate for meeting 10 CFR Part 37 requirements?

A1: The regulation was designed to allow licensees flexibility to choose methods that work best in each licensee's specific circumstances; therefore, the adequacy of a system will depend on the system's capabilities and site-specific requirements. It is the licensee's responsibility to choose monitoring and detection methods and technologies for effective compliance with the requirements of this section. The licensee can accomplish effective detection and monitoring in different ways tailored to different facility-specific operating conditions. The licensee's choices must also ensure that it has dependable means in place to transmit information between and among the various components used to detect and identify an unauthorized intrusion, to inform the licensee security staff, and to summon the appropriate responder.

Q2: May I use a less-sensitive monitoring system during the routine workday? May I rely on visual inspection by approved individuals during the normal workday and a system depending on an offsite monitoring facility at night?

A2: The licensee can meet detection and monitoring requirements by using electronic devices, visual monitoring, or a combination of these methods. However, the chosen method(s) must be effective in providing the required capability of immediate detection, assessment, and response. Personnel must be trustworthy and reliable, trained in appropriate security procedures, equipped for reliable communications, and capable of meeting the immediate detection requirements. Electronic devices must be capable of alerting trained onsite or offsite facility personnel upon first detection of an intrusion. Visual monitoring must also be capable of alerting trained assessment and response personnel without delay. For additional technical guidance on the capabilities and applications of different IDS technologies, the licensee may refer to NUREG-1959.

Q3: If a licensee uses electronic intrusion detection alarms, what minimum performance should these devices be capable of?

A3: To meet the requirements of 10 CFR 37.49(a)(2)(ii), a licensee must use intrusion detection alarms that will alert nearby facility personnel of unauthorized access to the security zone. These alarms may be visual (e.g., strobe lights); however, if the licensee relies on audible alarms to fulfill this function, the sound of these alarms should be distinguishable from other alarms and should be at least 65 decibels above ambient background noise level at the farthest location of a responder. (See DOE M 5632.1C-1, "Manual for Protection and Control of Safeguards and Security Interests," dated July 15, 1994, page 55.) If the licensee decides to use magnetic switches on doors and windows as part of an intrusion alarm system, the NRC considers the use of balanced magnetic switches a best practice because these switches are more difficult for an adversary to defeat. If the licensee uses motion detectors, it must use enough detectors to cover all potential entrance and egress points for radioactive material from the security zone. To meet the intent of 10 CFR 37.49(a), the detection system must be capable of detecting *all* unauthorized access to the security zone, including breaches of barriers used to isolate and control access to the protected radioactive material. The licensee's security plan should explain how any method used to alert facility personnel will be reliable and effective in permitting immediate detection on a continuous basis. For additional technical guidance on the capabilities and applications of different IDS technologies, the licensee may refer to NUREG-1959.

Q4: May I use silent alarms, as some experts recommend, for duress situations?

A4: The licensee may use silent alarms to meet the requirement in 10 CFR 37.49(a)(2)(i) for a monitored IDS linked to an onsite or offsite central monitoring facility. The licensee may deploy silent alarms in the immediate vicinity of the security zone in addition to the intrusion detection alarms to respond to situations in which anyone threatens or assaults another approved individual once inside the security zone. Silent alarms for such duress situations can give responders the advantage of surprise and can provide an additional increment of time to assess and more effectively intervene. However, if the licensee does choose silent alarms to comply with 10 CFR 37.49(a)(2)(ii), it must continuously monitor them by direct visual surveillance or a monitored video or audio surveillance system connected to a central monitoring station.

Q5: May I use radiation safety monitoring and detection systems for material security purposes?

A5: Yes. The licensee may use these systems if their use does not adversely affect radiation safety. To comply with this part, the licensee may use, for example, any monitored motion detection systems or alarms used to control access to high-radiation areas or other alerting systems used for radiation protection. The licensee may also modify these systems but only if the modifications do not compromise the equipment's original safety purpose. For example, the licensee should not relocate an integrated motion detection and alarm system originally installed at one location for radiation safety purposes to serve a security zone if the relocation would diminish the system's sensitivity at its original location. Radiation detection alarms would generally not enable the licensee to detect intrusions as required by 10 CFR 37.49(a)(2)(ii; however, the licensee may use these alarms to comply with the requirement in 10 CFR 37.49(a)(3) to detect the unauthorized removal of radioactive material from a security zone. The licensee's security plan should describe how any radiation safety systems used for security purposes will provide the required intrusion or removal detection without impairing these systems' performance or functionality for radiation protection.

§ 37.49, "Monitoring, Detection, and Assessment" (continued)

§ 37.49(a)(3)

A licensee subject to this subpart shall also have a means to detect unauthorized removal of the radioactive material from the security zone. This detection capability must provide:

§ 37.49(a)(3)(i)

For category 1 quantities of radioactive material, immediate detection of any attempted unauthorized removal of the radioactive material from the security zone. Such immediate detection capability must be provided by:

§ 37.49(a)(3)(i)(A)

Electronic sensors linked to an alarm; or

§ 37.49(a)(3)(i)(B)

Continuous monitored video surveillance; or

§ 37.49(a)(3)(i)(C)

Direct visual surveillance.

§ 37.49(a)(3)(ii)

For category 2 quantities of radioactive material, weekly verification through physical checks, tamper indicating devices, use, or other means to ensure that the radioactive material is present.

Explanation

Licensees with category 1 quantities of material must provide immediate detection capability through electronic sensors linked to an alarm, continuous monitored video surveillance, or direct visual surveillance. Licensees with category 2 quantities must ensure that the material is present through use, tamper-indicating devices, physical checks, or other means.

Q&As

Q1: What does NRC mean by the requirement in 10 CFR 37.49(a)(3)(i) that my detection capability be "immediate"?

A1: The common dictionary meaning of "immediately" is "at once, instantly, without any intervening time." Thus, the licensee's detection system must be capable of alerting it of any attempted unauthorized removal of radioactive material from the security zone instantly and without delay.

Q2: To meet the additional requirement of 10 CFR 37.49(a)(3)(i) for immediate detection of an attempted removal of a category 1 quantity of material from a security zone, can a licensee rely only on its main sitewide intrusion detection system linked to a monitoring facility?

A2: No. This requirement is in addition to the requirement to detect, assess, and respond to unauthorized access to the security zone. Methods that the licensee may use to meet the requirement to detect removal include, but are not limited, to the following:

- an alarming, electronic, tamper-indicating device
- radiation detectors that will alarm when the material is removed from a security zone
- continuous visual surveillance by an approved individual

If a licensee uses electronic, tamper-indicating alarms, the alarm should be capable of alarming either when an attempt is made to remove a category 1 quantity of radioactive material from a device or when an attempt is made to remove the device itself. The tamper-indicating alarms should be armed at all times, except during periods of maintenance.

If a licensee decides to use alarms that will alert nearby facility personnel to unauthorized access to the security zone, the audible alarm should be distinguishable from other alarms and should be at least 65 decibels above ambient background noise level at the farthest location of a responder. (See page 55 of DOE M 5632.1C-1.) The licensee's security plan should explain how any method used to detect unauthorized removal will be reliable and effective in providing immediate detection on a continuous basis.

Q3: Would licensees with continuous staffing on site, 24 hours a day, 7 days a week still be required to implement 10 CFR Part 37 requirements for immediate detection of a removal?

A3: Licensees will still be required to implement 10 CFR Part 37 requirements; however, they may be able to take credit for staffing, 24 hours a day, 7 days a week, to meet some of these requirements. The licensee can consider having staff members who can identify anyone without a clear work-related duty to be near a protected device and who will immediately call for assistance when needed as part of its response to the unauthorized removal of radioactive material.

Q4: May I use approved individuals to monitor for unauthorized removals of category 1 material if these individuals are moving about constantly and may not always be watching the area under access controls?

A4: The licensee may use any of a number of methods to detect an unauthorized removal; however, the method that it selects must meet the 10 CFR Part 37 requirements to respond immediately to any actual or attempted theft, sabotage, or diversion of protected radioactive material. Thus, if the licensee relies upon approved individuals to detect removal immediately, as required by 10 CFR 37.49(a)(3)(i), at least one individual must be observing at all times, unless another approved individual is escorting an unapproved individual in the affected security zone. However, no one method of monitoring may be right for all licensees.

Q5: Would periodic checking by a trained security guard meet the requirement for "immediate detection" of an actual or attempted removal of a category 1 quantity? Why would 15 minutes not permit an adequate response to an alarm?

A5: As noted earlier, "immediate" is commonly defined as "instant" or "without delay."

Although trained individuals can be used to monitor and immediately detect, assess, and respond as required, periodic checks, even those at intervals shorter than 15 minutes, will not meet the immediate detection requirement. Immediate detection is essential to reduce the risk that an adversary will be able to remove the material even if he or she is able to gain unauthorized access to it. Immediate detection is also essential to enhance the likelihood of recovering the material before it can be misused or to mitigate the consequences of misuse even if the adversary is able to remove it from its authorized location.

Q6: Is there any way to monitor without an alarm system?

A6: Trained individuals may be used to fulfill some requirements in 10 CFR Part 37; however, to comply with the requirement in 10 CFR 37.49(a)(1) to "maintain the capability to continuously monitor and detect without delay all unauthorized entries into...security zones," a licensee must maintain this capability 24 hours a day, 7 days a week. An appropriate method of monitoring, detection, and assessment should also provide a dependable means to transmit information between the various components used to detect an unauthorized intrusion, to inform security staff, and to summon the appropriate responder. An integrated alarm system may be the most cost-effective method to provide such continuous compliance.

Q7: Can a partial exemption from 10 CFR 37.49 be granted to licensees at extremely remote locations, such as offshore and wilderness sites, where access is limited and communication difficult?

A7: Any licensee may request an exemption from any NRC requirement; however, a licensee seeking an exemption from immediate detection requirements solely on the basis of remoteness would need to address the potential security risks of such relief. Remoteness may allow an intruder to gain undetected and unauthorized access more readily than in more populated environments. An additional lag in communications about an unauthorized removal may also provide advantages to an intruder's planning to steal or sabotage material. Licensees must therefore meet all applicable 10 CFR Part 37 requirements unless the NRC approves an exemption request. If a licensee does request an exemption from a 10 CFR Part 37 requirement because of remoteness or the difficulty of timely communication, the licensee must justify the request and should propose compensatory measures to mitigate the consequences of not meeting the specified requirement. The NRC will evaluate exemption requests on a case-by-case basis. See the Q&As for 10 CFR 37.11.

Q8: Could radiation detection equipment connected to a silent alarm be used to alert local law enforcement of an attempted theft of radioactive material?

A8: Yes. Depending on the configuration of shielding surrounding the radioactive source and other factors, radiation meters could be a means to detect and alert an LLEA of an attempted theft of radioactive material. However, a licensee would be prudent to avoid overreliance on any single method for detecting an unauthorized removal of radioactive material. The NRC designed the requirements for 10 CFR Part 37 to provide a defense-in-depth strategy, and the NRC expects licensees to consider all credible scenarios when developing and implementing a security program to implement the requirements in 10 CFR Part 37.

Q9: If a licensee with a category 2 quantity of radioactive material does not use the material weekly or make weekly physical checks to verify its continuing presence, what "other means" may this licensee use to detect the removal of a category 2 quantity of radioactive material under 10 CFR 37.49(a)(3)(ii)?

A9: The intent of the provision for "other means" to detect the removal of a category 2 quantity of material is to give the licensee flexibility to use the method that works best for its facility. Although electronic sensors for detecting the removal of a category 2 quantity of material are not required, the licensee should consider, as good practice, the application of these devices to category 2 quantities, where feasible, for immediate detection capability. For additional technical guidance on the capabilities and applications of different intrusion detection system technologies, the licensee may refer to NUREG-1959.

If a licensee decides to use an electronic, tamper-indicating device for detecting the removal of a category 2 quantity of material, the NRC recommends that the system be designed to silently or audibly alarm on any attempt to remove a device or a source from its device. The licensee should arm the tamper-indicating alarm at all times, other than those for equipment maintenance or calibration. The licensee's security plan should also explain how any method used to detect the unauthorized removal of a category 2 quantity of material will be reliable and effective in meeting the requirements in 10 CFR 37.49(a)(3).

§ 37.49, "Monitoring, Detection, and Assessment" (continued)

§ 37.49(b), "Assessment"

Licensees shall immediately assess each actual or attempted unauthorized entry into the security zone to determine whether the unauthorized access was an actual or attempted theft, sabotage, or diversion.

Explanation

A licensee must assess each actual or attempted unauthorized entry into a security zone immediately to determine if there was an actual or attempted theft, sabotage, or diversion.

Q&As

Q1: Can a licensee use automated devices to assess an intrusion and alert an LLEA?

A1: Yes. The licensee can use these devices if it is able to meet the requirements in 10 CFR Part 37. Assessment may be done by either automated devices or trained personnel who can initiate the appropriate response; however, in either case, the assessment must enable the licensee to request assistance and to begin any other mitigating measures immediately. Depending on the security system, the layout of security zones, and the design capabilities of the sensors, automated devices or systems may be programmed to summon LLEA assistance automatically in response to an intrusion alarm. The security plan and implementing procedures must describe how the licensee would assess and respond to unauthorized access. In developing its plan and procedures, the licensee should consider the possibility of simultaneous alarms at multiple locations.

Q2: Can licensees perform their own vulnerability assessment and change Part 37 time requirements for detection and response?

A2: Licensees are welcome to conduct vulnerability assessments as a good practice; however, they may not change the rule's time requirements for detection and response unilaterally. The regulations in 10 CFR Part 37 require licensees to detect, assess, and respond to any unauthorized access to the security zone "without delay," and the LLEA must be notified as soon as possible. If, for some reason, a licensee cannot comply with a specific requirement or if the licensee believes that the requirement is counterproductive, it should inform the NRC and request an exemption. The exemption request should propose compensatory measures to mitigate the consequences of not meeting the specified requirement.

Q3: Must I summon LLEA assistance in response to *any* intrusion alarm?

A3: No. The licensee must assess the situation immediately to determine if the alarm signals an actual or attempted intrusion and must decide if the intrusion justifies LLEA involvement. However, in certain instances, the licensee may determine in advance that an alarm from an electronic, tamper-indicating device calls for an automatic summons to the LLEA; however, not summoning an LLEA for every unauthorized intrusion would be good practice. Such intrusions could be either unknowing or inadvertent, and it would be counterproductive to request LLEA assistance automatically for such insignificant security events as an inadvertent activation of an alarm. Frequent alerts to the licensee's LLEA(s) are likely to result in a delayed or inadequate response to a real security incident.

§ 37.49, "Monitoring, Detection, and Assessment" (continued)

§ 37.49(c), "Personnel Communications and Data Transmission"

For personnel and automated or electronic systems supporting the licensee's monitoring, detection, and assessment systems, licensees shall:

§ 37.49(c)(1)

Maintain continuous capability for personnel communication and electronic data transmission and processing among site security systems; and

§ 37.49(c)(2)

Provide an alternative communication capability for personnel, and an alternative data transmission and processing capability, in the event of a loss of the primary means of communication or data transmission and processing. Alternative communications and data transmission systems may not be subject to the same failure modes as the primary systems.

Explanation

Licensees must maintain continuous capability for personnel communication and electronic data transmission and processing among site security systems. Licensees must also provide alternative personnel communication and data transmission and processing capabilities in case the primary means is lost. Alternative systems may not have the same failure modes as primary systems.

Q&As

Q1: What do I have to do to "maintain continuous capability" under 10 CFR 37.49(c)(1)??

A1: The licensee must have a dependable means to transmit information to all the various components involved in the detection and assessment of an intrusion, including the appropriate responder, 24 hours a day, 7 days a week, 365 days a year. The licensee may use land-line telephones, automatic dialers, cellular phones, pagers, radios, and other similar modes of communication to fulfill this requirement. When using more than one person for detection and assessment, the licensee must also provide a means for the various monitoring personnel to communicate with each other.

Q2: What personnel communications and data transmission systems are subject to the requirement for an alternative capability?

A2: The licensee must have an alternative capability for any primary system of communication or data transmission and processing that it relies upon to meet requirements. The alternative means must be able to provide continuous communication or data transmission capability. The licensee could use land-line phones, automatic dialers, cellular phones, pagers, radios, and other similar modes of communication to fulfill this requirement as long as they are not subject to the same failure mode as the primary systems that they must replace. For example, a radio or cellular phone could be considered as a backup to a land-line phone. However, an alternative cell phone system may not rely on the same cell tower and transmission system as the primary cell phone system.

Q3: To comply with the requirement that the alternative communication system not be subject to the same failure mode, may a licensee use a different cell phone service as a backup to a primary cell phone service?

A3: Yes. However, the licensee will need to show that the alternative cell phone service does not use the same satellite communications system, signal processing, or receiving tower as the primary service.

§ 37.49, "Monitoring, Detection, and Assessment" (continued)

§ 37.49(d), "Response"

Licensees shall immediately respond to any actual or attempted unauthorized access to the security zones, or actual or attempted theft, sabotage, or diversion of category 1 or category 2 quantities of radioactive material at licensee facilities or temporary jobsites. For any unauthorized access involving an actual or attempted theft, sabotage, or diversion of category 1 or category 2 quantities of radioactive material, the licensee's response shall include requesting, without delay, an armed response from the LLEA.

Explanation

Licensees must respond immediately to any actual or attempted theft, sabotage, or diversion of category 1 or category 2 material or unauthorized access to the security zones at licensee facilities or temporary jobsites. The licensee must request an armed LLEA response without delay.

Q&As

Q1: What do I need to do if I detect an intrusion into a security zone?

A1: The licensee's response would depend on its assessment of the purpose of the Intrusion; however, a response is required without delay. If the licensee assesses that the intrusion is an actual or attempted theft, sabotage, or diversion of category 1 or category 2 quantities of radioactive material, it must notify and request an armed response from the appropriate LLEA immediately, followed soon after by a call to the NRC Operations Center at (301) 816-5100. An immediate response by the licensee would permit a more timely response from law enforcement, thereby reducing the consequences of the incident.

The licensee's decision to call the LLEA and the NRC would depend not only on its assessment of the intent of the unauthorized access but also on if the area in which the breach occurred is an area that it had previously determined was in need of monitoring to meet the NRC's physical protection requirements. Thus, the licensee's assessment and response to an intrusion alarm in the business office section of its facility could be entirely different from its assessment and response to an intrusion alarm in a radioactive material storage area if the two areas are not adjacent.

Q2: Could radiation detection meters connected to a silent alarm be used to alert local law enforcement of an attempted theft of radioactive material?

A2: Yes. The licensee could use radiation meters as part of a system to alert an affected LLEA of an actual or attempted theft of radioactive material if this method is acceptable to the LLEA. The rule is designed to provide a defense-in-depth strategy for radioactive material security, and the NRC expects licensees to consider other measures for requesting LLEA assistance if such assistance would also be necessary.

Q3: Since licensed facilities are also broken into to obtain equipment items other than radioactive materials, would coordination require licensees to summon the LLEA every time there is a break-in?

A3: No. The licensee's decision on if it should call the LLEA would depend on what areas, as determined by the licensee, need to be controlled for access to the radioactive material and would depend on the licensee's assessment of the intent of the unauthorized access. A licensee's assessment and response to an intrusion alarm in a conference room or office supply storage area could be entirely different from its assessment and response to an intrusion alarm in a security zone for a radioactive materials storage area.

Q4: Do I have to conduct response exercises with my LLEA?

A4: No. Exercising the response portion of a licensee's security plan is a good practice; however, it is not required. Some LLEAs may not be willing or able to participate, and the NRC also recognizes that requiring all licensees to exercise their response plans may not be cost effective for small licensees with less complex security plans. However, licensees' security programs may benefit from exercising their response plans with their affected LLEA(s).

§ 37.51, "Maintenance and Testing"

§ 37.51(a)

Each licensee subject to this subpart shall implement a maintenance and testing program to ensure that intrusion alarms, associated communication systems, and other physical components of the systems used to secure or detect unauthorized access to radioactive material are maintained in operable condition and are capable of performing their intended function when needed. The equipment relied on to meet the security requirements of this part must be inspected and tested for operability and performance at the manufacturer's suggested frequency. If there is no manufacturer's suggested frequency, the testing must be performed at least annually, not to exceed 12 months.

Explanation

A licensee must implement a maintenance and testing program to ensure that intrusion alarms, associated communication systems, and other physical components of their detection systems remain operable and can perform their intended function when needed. Inspection and testing must be performed at the manufacturer's suggested frequency or at least annually if the manufacturer has no suggested frequency.

Q&As

Q1: What minimum measures should a licensee implement for a maintenance and testing program?

A1: A licensee should do the following:

- Identify all alarms, communication systems, and other physical components necessary to secure radioactive materials or to detect unauthorized access to them.
- Specify the intended function of each component identified in the program and the minimum performance required to fulfill that function.
- Specify the test(s) that it will conduct on each component and identify the minimum quantitative or qualitative test results that are required for finding the component operable and capable of performing its intended function.
- Identify the testing equipment that it will use and prescribe any device-specific procedures necessary for the use or maintenance of this equipment.
- Identify the measures that it will apply to ensure that the testing equipment used in the program will perform inservice as expected.
- Prescribe procedures for the routine maintenance of each intrusion alarm, communications system, and physical component of both the system used to secure the subject radioactive material and the system used to detect unauthorized access.
- Require a written record for each test and maintenance activity performed on the security or detection system.

Q2: Do I have to inspect and test my facility's security systems every month or every quarter year to meet this requirement?

A2: No. The regulation in 10 CFR 37.51(a) requires the licensee to conduct inspection and testing for operability and performance needs at the manufacturer's suggested frequency. If the manufacturer has no suggested frequency, the licensee must perform the testing at least annually, not to exceed 12 months. However, regular daily or weekly use may confirm the operability and performance of some components, such as personnel and data communications systems.

Q3: Must I test and maintain every intrusion detection system or other security system on my facility site? If I determine that a detection or security device is unnecessary to meet 10 CFR Part 37 requirements, but still keep it energized, must I keep testing and maintaining it?

A3: No. The licensee is required to test and maintain only those components that it relies upon to meet the security requirements of this part for Category 2 or greater quantities of radioactive material. Because the regulation in 10 CFR 37.49(c)(2) also requires the licensee to provide alternative communication capabilities for transmitting personnel and data and for processing in the event of a loss of the primary capability, it must also test and maintain the systems for these alternative communication capabilities.

Q4: What's the difference between maintaining a system or component in operable condition and ensuring that it is capable of performing its intended function when needed? Would licensees have to administer different tests for performance than for operability?

A4: The need for different tests would depend on the intended function of the system or component and on if that function of the system or component could be fulfilled if the system or component were only in operable condition. For example, the licensee would generally not need to subject a locking deadbolt on a door that isolates radioactive materials to any additional testing as long as it could lock and unlock the door when necessary. However, an intrusion alarm might be considered "in operable condition" if it can be turned on and made to produce an audible sound, but it would not necessarily be "capable of performing its intended function" if it could not reliably detect the movement of an intruder with sufficient sensitivity at a given distance or if it could not produce a sound of sufficient volume to be audible at the nearest work area equipped to assess and respond.

Q5: What kinds of tests for intrusion alarms would be acceptable for demonstrating compliance?

A5: The licensee should refer to the manufacturer's recommended testing protocol, if one is available, for each system or piece of equipment subject to testing and maintenance under 10 CFR 37.51. For additional guidance on IDS testing and maintenance, the licensee may also refer to Chapters 8.6 and 8.7 of NUREG-1959. Chapters 9.4 and 9.5 of this NUREG report also cover testing and maintenance for backup power sources.

Q6: What should a licensee consider to be an "associated communication system" subject to testing and maintenance? Would a closed-circuit television monitoring system be considered a communication system, even if it's intended primarily for the timely detection of an intrusion?

A6: The NRC considers a communications system to be "associated" if it fulfills any function essential for the operability or performance of the security or detection system as a whole, not just for the performance of intrusion alarms. For the purposes of this section, the NRC considers a "communication system" to be any device or network of devices used to transmit voice, video, data, or other information from a person or machine at one location to a recipient person or machine at another. This device could include a land-line phone, a walkie-talkie system, or a cell phone. A closed-circuit television monitoring system could also be considered a communication system within the meaning of this definition even if the system also performs monitoring or detection functions. An alarm system that incorporates motion-sensing devices could also be considered a communications system even if it also performs monitoring and detection functions. However, in either case, the testing and maintenance requirements in this section would apply because 10 CFR 37.51 applies to systems for securing radioactive materials and to systems for detecting unauthorized access to these materials. Any system designed to perform multiple functions would be subject to this section as long as one of the system's functions supported either security or detection. For this example, the communication system comprises the lines or cabling that allows transmission of the data from the closed-circuit television cameras to the recipient monitoring station. Such data lines or cabling is also subject to testing and maintenance.

Q7: Specifically, what "other physical components of the system" used for securing or monitoring access to radioactive material would be subject to the requirements for testing and maintenance? Aren't all "components" of these systems inherently "physical"?

A7: To enable licensees to identify physical components of security systems other than alarms and communications systems, the NRC is using the word "physical" to distinguish tangible material components, such as architectural structures and equipment, from system "components" that comprise human beings and their actions, including the plans and procedures that govern those actions. Thus, for a system that secures category 1 or category 2 quantities of radioactive material from unauthorized access within a security zone under 10 CFR 37.47(c)(1), any other component should be considered an "other physical component" if it meets the following two criteria:

1. It is not otherwise integrated into an alarm or communications system, nor is it an employee or contractor performing security-related work at the licensee's site.

2. Its intended function supports the isolation of these radioactive materials by the use of continuous physical barriers that allow access only through established access control points.

Examples of "other physical components" of security systems may thus include walls, doors, remotely operated doors, ceilings, floors, windows, storage containers, shielding, scales, mounting bolts, fasteners, key card systems, locks, keys (if applicable), emergency alternate-power generators, and lighting.

Correspondingly, a component of a system used to maintain the capability to continuously monitor and detect, without delay, all unauthorized entries into its security zones under 10 CFR 37.49(a)(1) should be considered an "other physical component" if it meets the following two criteria:

1. It is not otherwise integrated into an alarm or communications system, nor is it an employee or contractor performing security-related work at the licensee's site.
2. Its intended function supports the continuous monitoring and immediate detection of all unauthorized entries into the licensee's security zone(s).

Examples of "other physical components" of detection systems may include video surveillance cameras and monitors, night-vision devices, motion sensors for self-illuminating floodlights, and tamper-indicating devices that are not otherwise integrated into an alarm or communication system. If a licensee opts to provide an alarm system for a loss of the primary power source for monitoring and detection systems, that system is also an example of an "other physical component" subject to testing and maintenance. If the licensee chooses instead to use an alternate- or auxiliary-power source, such as a gasoline-fueled generator, that alternate-power source would also be subject to testing and maintenance. See Q2 and A2 for 10 CFR 37.49(a)(1) above. The NRC recognizes that not all physical barriers will need to be "tested" in the sense of the term commonly understood to involve a rigorous, documented administration of specified evaluation procedures under controlled and often nonroutine conditions. The purpose of testing in this regulatory context is to confirm that the subject system component is operable and is capable of performing its intended function when necessary.

The licensee can use a number of methods to obtain confirmation that the subject system component is operable and is capable of performing its intended function when necessary; not all of these methods require the quantitative methods often associated with instrumentation and testing protocols. The operability of some physical barriers, such as keys, key cards, doors, walls, windows, two-way mirrors, and floors, is regularly and satisfactorily confirmed in the course of routine operations. The NRC recognizes that the licensee may not be able to test the performance capability of many physical barrier system components, such as tamper-indicating devices and jersey barriers, as security barriers without risking or resulting in a degradation of their performance. Other barriers, such as the locked entrance to a panoramic irradiator, serve a safety function for which testing could complicate the ALARA principle of radiation safety. In these cases, the NRC does not expect a licensee to conduct testing for a physical barrier's performance if the test itself could compromise either radiation safety or the future performance of the component or system.

Q8: Will testing and maintenance have to be performed by approved individuals meeting the access authorization requirements of Subpart B?

A8: No. However, individuals who are not approved for unescorted access will need to be escorted when they are performing maintenance or testing activities within a security zone.

Q9: Will testing equipment need to be secured within a security zone?

A9: No. If the testing equipment is susceptible to tampering by an insider, keeping this equipment within the security zone could be a practicable way for some licensees to control unauthorized access to this equipment; however, the NRC recognizes that, for space, operational, and other considerations, storage of security-sensitive testing equipment inside a security zone may not be feasible. In such cases, the licensee should consider other ways to control access to this equipment, such as a locker with a key pad, controlled key, key card, or combination lock.

Q10: What do I have to do to have an acceptable testing and maintenance program for my intrusion detection system (IDS)?

A10: The regulation in 10 CFR 37.51(a) requires the licensee's program to cover the testing and maintenance of all IDSs that it relies upon to detect unauthorized access to radioactive material. The program must ensure that these systems and components are maintained in operable condition and are capable of performing their intended function when necessary. To maintain functionality, the licensee must periodically test the IDS and must perform maintenance on malfunctioning components. The NRC considers the testing program acceptable if the IDS operates in a manner consistent with the licensee's physical security plan. The licensee must test the entire IDS or components of the IDS at the frequency specified by the manufacturer or at least annually. The licensee may choose to test the entire IDS or components of the IDS throughout the 12 months of the testing cycle.

§ 37.51, "Maintenance and Testing" (continued)

§ 37.51(b)

The licensee shall maintain records on the maintenance and testing activities for 3 years.

Explanation

Maintenance and testing records must be kept for 3 years.

Q&As

Q1: What kinds of records of maintenance and testing activities should a licensee maintain?

A1: For each maintenance activity, a record should identify the following items:

- the name(s) of the person(s) who performed the maintenance
- the date that the maintenance was performed

- the component(s) or system(s) on which the maintenance was performed
- the purpose of the maintenance, identifying, as appropriate, the deficiencies in operability or performance
- any maintenance activities needed to remove any deficiency in the operability or performance of the component or system

For each testing activity, a record should identify the following items:

- the name(s) of the person(s) who performed the testing
- the date of the testing
- the component(s) or system(s) tested
- the purpose of the testing
- the performance expected to fulfill the component's or system's intended function
- the results of the testing
- any maintenance activities needed to remove any deficiency in the operability or performance of the component or system

§ 37.53, "Requirements for Mobile Devices"

§ 37.53

Each licensee that possesses mobile devices containing category 1 or category 2 quantities of radioactive material must:

§ 37.53(a)

Have two independent physical controls that form tangible barriers to secure the material from unauthorized removal when the device is not under direct control and constant surveillance by the licensee.

Explanation

Mobile devices containing category 1 or category 2 quantities of radioactive material must have two independent tangible barriers to secure the material when the device is not under direct control and constant surveillance.

Q&As

Q1: What is the performance objective of the requirement for two independent physical controls against the unauthorized removal of radioactive material when it's not under direct control and constant surveillance?

A1: Because mobile devices can often be easily concealed and transported, they are particularly vulnerable to attempted theft or diversion. A mobile device could even be removed before the licensee has an opportunity to respond to an intrusion. Therefore, the objective of this requirement is to delay an attempt to steal the device by making it harder to steal. Achieving this objective would provide additional time for the licensee to detect, assess, and respond to an attempt by notifying an LLEA if necessary.

Q2: What does the NRC mean by "independent" physical controls?

A2: For purposes of compliance with this requirement, an "independent" physical control does not rely on any other system to deter or delay unauthorized removal of the radioactive material when the mobile device containing it is not under the licensee's direct control and constant surveillance. Such a physical control must provide a barrier against unauthorized removal, regardless if any other barrier or system is disabled.

Examples of three independent physical controls used for category 1 or category 2 quantities of radioactive material in a mobile device at a licensed facility are as follows:

1. storage of the device inside a locked storage shed within a secured outdoor area, such as a fenced parking area with a locked gate
2. storage of the device in a room with a locked door within a secured building for which access is controlled by a separate lock and key or by a security guard
3. storage of the device inside a locked, nonportable cabinet inside a room with a locked door if the building is not secured

Examples of two independent physical controls used when the licensee secures the radioactive material in or on a transportation vehicle are as follows:

1. storage of the mobile device in a container that is physically attached to a vehicle with the container secured by two separate chains or steel cables, each of which is locked and separately attached to the vehicle in such a manner that the container cannot be opened without the removal of both of the chains or cables
2. storage of the device in a container inside a locked trunk, camper shell, van, or other similar enclosure with the container physically secured to the vehicle by a locked chain or steel cable in such a manner that the container could not be opened and the mobile device removed without both breaking into the enclosure on the vehicle and removing the chain or cable

Examples of two independent physical controls used at a temporary jobsite or at a location other than a licensed facility or a licensee's vehicle are as follows:

1. storage of the mobile device inside a locked building in a locked nonportable structure with the device physically secured by a locked chain or steel cable to the nonportable structure in such a manner that the device could not be removed without removing the chain or cable[2]
2. storage of the device in a locked garage within a locked vehicle or physically secured by a locked chain or steel cable to a disabled vehicle in such a manner that the device could not be removed without removing the chain or cable

[2] Note that a construction trailer, sea container, or parked and locked truck should not be considered a nonportable structure. See the Q&As below and those for the definition of "mobile device" in 10 CFR 37.5.

For large mobile devices on wheels that are used inside a facility, a variety of other independent physical controls, including the following, may accomplish additional delay:

- speed bumps too large for the device to traverse on the facility floor
- elevated doorway thresholds
- protective storage enclosures
- channels in a floor large enough to catch the device's wheels
- fixed or retractable bollards
- wheel locks, including steering wheel locks (made of hardened material) that require a key or special tool to release
- a hardened chain and lock that cannot be easily cut

However, these additional physical controls for security purposes should not compromise safety. If improperly implemented, some of the suggested controls, such as elevated door thresholds and channels in a floor, may compromise occupational safety, and a licensee that intends to use such controls should address these issues.

Q3: Does this section require two physical barriers *in addition to* the physical barrier required for the security zone, or does a barrier for a security zone count as one of the two required barriers?

A3: A barrier for a security zone may count as one of the two required barriers if it is independent of the other barrier and has not been defeated (e.g., for lock replacement or the servicing of a key card device). Although the two barriers may be of the same type, such as a locked room within a locked room, the barriers have different purposes that contribute to defense-in-depth security. The tangible barriers required by 10 CFR 37.53 are to secure the material from unauthorized *removal* when the device is not under the licensee's direct control and constant surveillance. The security zones required by 10 CFR 37.47 must control unescorted *access*, and licensees may control unescorted access by using continuous physical barriers that allow access to a security zone only through established access control points. A physical barrier for a security zone could control access to radioactive material and secure it from unauthorized removal; however, it must serve both purposes simultaneously and may not rely upon any other measures, such as the remote monitoring of the security zone entrance, to fulfill both functions.

Q4: If mobile devices are stored in one room, does that mean the licensee has an aggregated quantity of radioactive materials?

A4: The regulation provides that licensees must consider radioactive materials aggregated if the following criteria apply:

- Their total quantity at a single location equals or exceeds a category 2 quantity using the sum-of-fractions method. See the Q&As on the definition of "aggregated" in 10 CFR 37.5.
- Breaching a common physical security barrier (e.g., a locked door at the entrance to a storage room) would allow access to the radioactive material or devices that contain the radioactive material. See the Q&As on physical barriers in 10 CFR 37.47.

For example, if the licensee stores several mobile devices—each of which is below the category 2 quantity threshold—in one room without any physical barriers between them, these devices would be considered aggregated if their combined activity equaled or exceeded the category 2 threshold.

Q5: If I have a category 1 or 2 quantity of radioactive material in a container that isn't mounted on wheels or casters and can't be carried by hand, and I secure it in a pickup truck bed or on a trailer, would the NRC consider this source or device "mobile"?

A5: Yes. The NRC would consider the source or device mobile because the device would be "mounted on wheels" (i.e., the truck/trailer wheels) and "equipped for moving without a need for disassembly or dismounting" under the definition of "mobile device." A container or device on a pickup truck bed or on a trailer would also subject the truck or tractor to the requirement in 10 CFR 37.53(b) for disabling the vehicle when the licensee does not have the container or device under direct control and constant surveillance. See the Q&As on 10 CFR 37.53(b).

Q6: If I have a radiography camera or other mobile device with a category 1 or 2 quantity of radioactive material, under what conditions would that device meet the exclusion for "stationary equipment installed in a fixed location" under the definition of "mobile device"?

A6: Under this definition, the device would need to be welded, bolted, or otherwise affixed to an immovable piece of equipment, such as an assembly line, in a manner that would require the disassembly and dismounting of additional equipment before the device could be moved. The device could also be excluded as "stationary equipment installed in a fixed location" if it were chained or cabled to a stationary piece of equipment and secured with a lock; however, the licensee would need to control access to the combination, key, password, or other means of unlocking the lock.

§ 37.53, "Requirements for Mobile Devices" (continued)

§ 37.53(b)

For devices in or on a vehicle or trailer, unless the health and safety requirements for a site prohibit the disabling of the vehicle, the licensee shall utilize a method to disable the vehicle or trailer when not under direct control and constant surveillance by the licensee. Licensees shall not rely on the removal of an ignition key to meet this requirement.

Explanation

Unless health and safety requirements for a site prohibit the disabling of vehicles, the licensee must disable them when they are not under direct control and constant surveillance. Licensees may not rely on the removal of an ignition key to meet this requirement.

Q&As

Q1: What are acceptable methods to disable a vehicle or trailer when it is not under direct control and constant surveillance by the licensee?

A1: The objective of the vehicle-disabling requirement is to delay unauthorized removal of a device by stealing the vehicle on which it is secured. Examples of acceptable vehicle-disabling methods would include trailer hitch locks, wheel locks ("boots"), or methods to disable the vehicle's engine. The licensee cannot consider the removal of a standard key from a vehicle's ignition sufficient for disabling a vehicle's engine because a vehicle can be started without the key using, for example, a duplicated key or hot-wiring techniques. Many current advances in ignition and key technology may provide for additional barriers that would cause delay in accessing the radioactive material. An example is a key implanted with an electronic chip that is only recognizable to the computer programmed in the vehicle. Only this key, and not a duplicated key, would be able to start the vehicle. A licensee with this key system may request an exemption from 10 CFR 37.53(b) for that specific vehicle; however, the licensee would need to address the possibility of hot-wiring or other techniques to defeat the system. A similar disabling function may also be provided by biometric recognition systems when they become commercially available. Such technologies, which allow a vehicle to be operated only by a means that is not easily defeated, would be considered an appropriate means to disable a vehicle. However, the licensee should rely on deflating the tires as an effective immobilization technique because an adversary could easily re-inflate them with an air pump or pressurized gas cartridge.

Q2: When a device in or on a vehicle is not under the licensee's direct control and constant surveillance, this paragraph requires the licensee to disable the vehicle "unless health and safety requirements at a site "prohibit" such disabling. What kind of health and safety requirements at a site would the NRC deem to "prohibit" vehicle disabling?

A2: The facility or site requirements or procedures would need to specify that vehicles cannot be disabled or that they need to remain accessible for immediate relocation in an emergency. The regulation will not require the disabling of a vehicle or trailer at any site at which a known or reasonably foreseeable operational hazard, such as an explosion, fire, or release of a toxic substance, could require an individual to operate the vehicle immediately to escape or to mitigate the hazard.

Q3: Does the requirement to secure mobile devices mean that all trucks will need an alarm system when they carry a device containing a category 1 or category 2 quantity of radioactive material?

A3: No. However, if the truck is left unattended, the licensee must have a way to monitor and immediately detect, assess, and respond to an actual or attempted theft, sabotage, or diversion. An alarm system is an acceptable method.

Q4: Can 1/8-inch wire cables be used to secure a radiography camera instead of chains? Is the lock then more of a vulnerability than the cable?

A4: Yes. The licensee may use a heavy-duty twisted steel wire cable instead of a chain to secure mobile devices. Ideally, the wire should be thick enough so that it can only be removed with a heavy-duty cable cutter (i.e., thickness greater than a 1/8-inch or 2-millimeter No. 10 wire). However, any system is only as effective as its weakest component, and other components of the securing mechanism, including the lock, will need to have similar strength such that it too would require a heavy-duty bolt cutter for removal (typically, a tensile force of 2,000 pound-force and a shackle-cutting force test of 4,000 pound-force). Regulatory Guide 5.12, "General Use of Locks in the Protection and Control of Facilities and Special Nuclear Materials," issued November 1973, may provide some useful information. Other references include NUREG/CR-5929, "Locking Systems for Physical Protection and Control," issued November 1992; NUREG-0274, "Catalog of Physical Protection Equipment," Book 2, Volume III, "Entry Control Components," Part 5, issued June 1977;

Army Regulation 190-11, "Physical Security of Arms, Ammunition, and Explosives," dated November 15, 2006; and DOE M 5632.1C-1.

Q5: Can two or more barriers with separate locks that share the same key or lock combination qualify as "two independent physical controls" as stipulated in the requirement for securing mobile devices?

A5: Yes. Two or more barriers with separate locks that share the same key or lock combination could qualify as "two independent physical controls." Whether separate locks use the same or different keys or combinations is an aspect of the licensee's access control program and does not determine if two barriers can be considered as "two independent physical controls." Despite the number of keys or combinations used, the most important test for ensuring that two independent physical controls exist is that each barrier is separate from the other and that neither relies upon the other to perform its intended function. The same guidance applies when considering barriers to determine if material is aggregated.

An important aspect of a licensee's physical protection program is controlling access at the licensee's facility. If a key-based system is used, it is essential that the licensee distribute the keys only to personnel who have been granted unescorted access and who have a need for access to perform their assigned duties. The licensee must ensure that those individuals who are part of its physical protection program understand the importance of their roles and responsibilities for controlling access, especially if these individuals control combinations or keys that secure material.

§ 37.55, "Security Program Review"

§ 37.55(a)

Each licensee shall be responsible for the continuing effectiveness of the security program. Each licensee shall ensure that the security program is reviewed to confirm compliance with the requirements of this subpart and that comprehensive actions are taken to correct any noncompliance that is identified. The review must include the radioactive material security program content and implementation. Each licensee shall periodically (at least annually) review the security program content and implementation.

§ 37.55(b)

The results of the review, along with any recommendations, must be documented. Each review report must identify conditions that are adverse to the proper performance of the security program; the cause of the condition(s); and, when appropriate, recommend corrective actions and corrective actions taken. The licensee shall review the findings and take any additional corrective actions necessary to preclude repetition of the condition, including reassessment of the deficient areas where indicated.

§ 37.55(c)

The licensee shall maintain the review documentation for 3 years.

Explanation

Each licensee must review its security program's content and implementation at least annually. Each review report must identify adverse conditions and the causes of each condition. When appropriate, the report must recommend corrective actions and review corrective actions taken. The licensee must review the report's findings and must take any additional corrective actions and reassessments necessary to preclude recurrence of the adverse condition(s).

Q&As

Q1: How should a licensee evaluate the "effectiveness" of its security program to comply with NRC requirements for program reviews?

A1: A licensee should consider several things when evaluating the continuing effectiveness of its security program. Specifically, it should consider its ability to confirm compliance with all the applicable requirements in this subpart and to take comprehensive and effective actions to correct identified noncompliances. However, most importantly, the licensee should keep in mind that continuing effectiveness is not a static condition. Continuing improvements are an essential part of an effective program. To minimize the potential for false negatives, the licensee's program reviews must address each applicable requirement of this part. These reviews should identify adverse conditions, noncompliances, and root causes and should provide recommended corrective actions. The licensee should follow up on the implementation of these actions and reassess their impacts. Therefore, the hallmark of an effective program is not the absence of recorded adverse conditions or noncompliances. It is the documented evidence that the licensee has made diligent efforts to find problems and that it is continuing to reassess the effectiveness of its actions to prevent them from reoccurring.

Q2: What does the NRC mean by its requirement to review the security program content and implementation "at least annually"?

A2: Recognizing that some demands on a licensee's time and resources are beyond its control, the NRC will consider that the licensee is conducting a program review "at least annually" if it conducts the review each year at about the same time of year. The licensee will comply if it reviews its program at regular intervals not to exceed 12 months.

Q3: Must a licensee engage an outside contractor to conduct security program reviews?

A3: No. Although hiring an independent party to conduct security program reviews would be one way for licensees to demonstrate compliance, the regulation does not require it. However, the licensee should, to the extent possible, try to avoid a situation in which the implementers of the program and their supervisors are reviewing their own work. If the licensee has a large enough staff, it could establish a review team of approved individuals led by an individual, such as the RO for access authorization decisions, who works outside the management chain of the licensee's security staff. If the licensee conducts activities with category 1 or category 2 quantities of radioactive material at more than one location with a different security staff at each location, it could have the review team at one location review the program implemented by the staff of a different facility.

Q4: Does a licensee need to report to the NRC any noncompliance identified during a security program review?

A4: No. However, in accordance with the regulation, the licensee must "ensure...that comprehensive actions are taken to correct any noncompliance that is identified" by the review. The licensee must also document the corrective actions taken and "take any additional corrective actions...to preclude repetition of the condition." This documentation must be made available for NRC inspection.

Q5: What would the NRC consider a "condition adverse to the proper performance of the security program"?

A5: The NRC would consider, as an adverse condition, any occurrence or continuing state that degrades or could degrade, if it is not corrected, the ability of the physical security system to detect, or respond to, an actual or attempted theft, sabotage, or diversion of a category 1 or category 2 quantity of radioactive material without delay. An example of an adverse condition might be a breach in a physical barrier, a missing or misdirected motion detector, or a door that fails to latch and lock itself when closed. An adequate program review should never be limited to looking only for existing or imminent noncompliances. It should assess or reassess all conditions that may call into question the continuing effectiveness of the licensee's security program.

Q6: The regulation requires the report resulting from a program review to recommend corrective actions "when appropriate." When should a licensee recommend corrective actions?

A6: The rule does not specify particular conditions for which corrective action is required; however, a program review report should, as prudent practice, recommend at least one corrective action for each noncompliance or "condition adverse to the proper performance of the security program" identified as a result of the review.

Q7: What should a licensee consider as "review documentation" for the purposes of this subsection?

A7: The licensee should retain the approved final version of its annual review report and any attachments or enclosures related to that report. Related records should include the membership and leadership of the review team (if applicable), a description of the management approval process for the annual report (if applicable), root cause analyses for identified noncompliances or adverse conditions, recommended corrective actions, evaluations of the effectiveness of past corrective actions, and other documents that were considered in the review. Review documentation should also include minority views on issues in the report on which significant professional disagreement was present. The NRC does not expect a licensee to retain meeting records, the notes of each member of a review team, or rough drafts of its annual security program reviews.

Q8: This section introduces the term "radioactive material security program." Is there a difference between this term and the term "security program" for the purposes of this section?

A8: No. Both refer to the security program required by 10 CFR Part 37.

§ 37.57, "Reporting of Events"

§ 37.57(a)

The licensee shall immediately notify the LLEA after determining that an unauthorized entry resulted in an actual or attempted theft, sabotage, or diversion of a category 1 or category 2 quantity of radioactive material. As soon as possible after initiating a response, but not at the expense of causing delay or interfering with the LLEA response to the event, the licensee shall notify the NRC Operations Center ((301) 816-5100). In no case shall the notification to the NRC be later than 4 hours after the discovery of any attempted or actual theft, sabotage, or diversion.

Explanation

A licensee must notify the LLEA immediately after determining that an unauthorized entry resulted in an actual or attempted theft, sabotage, or diversion of radioactive material. The licensee must also notify the NRC Operations Center as soon as possible (not later than 4 hours after discovery) but not at the expense of causing delay or interfering with the LLEA response to the event.

Q&As

Q1: The regulation requires a licensee to notify the LLEA immediately after initiating "an appropriate response" to an actual or attempted theft, sabotage, or diversion of a category 1 or category 2 quantity of radioactive material. What kind of response by the licensee would the NRC consider "appropriate"? Under what circumstances, if any, would an armed licensee response before the arrival of the LLEA be considered appropriate?

A1: As soon as the licensee has determined that an actual or attempted theft, sabotage, or diversion is in progress, an appropriate licensee response pending the arrival of LLEA assistance would depend on the licensee's assessment of the intruder's intent and his or her ability to carry out theft, sabotage, or diversion. A single unarmed individual who has only made an unauthorized entry at the perimeter of a security zone may call for a different response than the intrusion of several armed individuals who had removed, or who were about to remove, a category 1 or category 2 quantity of radioactive material. The appropriateness of the licensee's response should also consider the applicability of the licensee's written procedures to the actual circumstances. For example, these procedures could require the licensee to consider the proximity of the intrusion to the target material, the quantity of material at risk of sabotage or unauthorized removal, the quality and quantity of any remaining physical barriers to the removal of the material or the escape of the intruders, and if the physical barriers are likely to delay the intruders sufficiently for the anticipated arrival of the LLEA. The regulation does not require a licensee's security staff to be armed for its response to be deemed appropriate. A decision to arm a facility's security personnel is solely up to the licensee and would depend on applicable State laws.

Q2: The regulation requires the licensee to notify the NRC Operations Center "as soon as possible" after initiating a response, "but not at the expense of causing delay or interfering with the LLEA response to the event...[and] in no case...later than 4 hours after its discovery." What would the NRC consider to be causes of delay or interferences with the LLEA's response sufficient to justify a licensee decision to postpone notifying the NRC?

A2: Any justification for postponing a notification to the NRC would depend on the circumstances of the event. Perhaps the most compelling justification would be a situation requiring a licensee to devote all available resources to restoring its facility to a safe condition or to protecting individuals from actual or threatened physical harm, as in the case of a fire or explosion, or from an armed hostage-taking event. An ongoing attempt to steal the material or use it for sabotage could also require the licensee to assist the LLEA before or after its arrival. A similar necessity for the licensee's assistance could be the threat of a bomb hidden somewhere on the facility site. Under such circumstances, the responding licensee staff could be justified in postponing notification of the NRC until the LLEA had determined that the licensee's assistance in the response is no longer required. However, if an employee of the licensee outside of its onsite security staff became aware of the emergency and was not immediately needed to help the security staff respond to the event, the NRC would expect that employee to notify his or her management and would expect the licensee to notify the agency as soon as possible. The same would be true if the licensee discovered an apparent theft of a category 1 or category 2 quantity of radioactive material after the fact; the NRC would expect the licensee to notify it immediately after notifying the LLEA and initiating measures to ascertain if the missing material was still on the premises. However, the NRC can envision none but the most extreme and unusual circumstances to justify delaying a notification of more than 4 hours after the discovery of any attempted or actual theft, sabotage, or diversion.

Q3: To avoid an NRC notice of violation after the event, would I have to obtain confirmation from the LLEA that a delay in the licensee's notification of the NRC was necessary to avoid delaying or interfering with the LLEA's response?

A3: No. However, the licensee is not prohibited from seeking the LLEA's endorsement of its decision to delay NRC notification. The NRC would take the LLEA's judgment into account; however, the agency's determination of compliance or noncompliance with this notification requirement would rest principally on an assessment of the specific facts and circumstances of the situation, not on the LLEA's judgment.

Q4: Should notification to the LLEA be made by a radiographer at a temporary jobsite if a device is stolen from the temporary jobsite?

A4: Yes. The licensee should contact the LLEA that can provide the most prompt and effective response. (See also the Q&As for 10 CFR 37.45(d) and 10 CFR 37.49(d).)

Q5: When I determine that an actual or attempted theft, sabotage, or diversion of radioactive materials has begun, what information must I report to what organization (the NRC or LLEA)?

A5: In the event of any actual or attempted theft, sabotage, or diversion of radioactive material protected under 10 CFR Part 37, the licensee must notify the LLEA immediately, followed soon afterward by a call to the NRC Operations Center at (301) 816-5100. Telephone calls to notify the NRC should be made as prompt as possible but not at the expense of causing delay or interfering with the LLEA's response to the event.

The licensee's notification to the LLEA should provide, at minimum, a basic "who, what, when, and where" description of the event. To facilitate a timely and effective LLEA response, the licensee should also answer any follow-on questions from the LLEA. For example, these questions could request information on landmarks at the affected facility's location; an estimate of the kinds and quantities of radioactive material that could be affected and a description of the type of container for the material (e.g., a stationary or a mobile device or another type of container); the number of intruders believed to be involved and available information about their weaponry, equipment, and apparent objectives; the location(s) of the unauthorized activity within the facility and the closest safe access point for incoming LLEA personnel; a description of the physical barrier system deployed to isolate the radioactive materials from unauthorized access; and a summary description of the licensee's other security measures and resources available to assist LLEA responders. The licensee may give information to the LLEA on its physical protection of radioactive materials without violating the information protection requirements in 10 CFR 37.43(d).

The licensee's notification to the NRC should provide, at minimum, a basic "who, what, when, and where" description of the event. Depending on the NRC's need for additional information, the licensee could also provide some of the information already provided to the LLEA. For example, this information could include the number of intruders believed to be involved and available information about their weaponry, equipment, and apparent objectives; a description of the physical barrier system deployed to isolate the radioactive materials from unauthorized access; and a summary description of the licensee's other security measures and resources available to assist responders. However, for either kind of agency notification, the licensee should consider the known and potential needs of the recipient agency for mission-related information and should address these issues, if possible, beforehand in its security plan and procedures. The licensee should also consider that compliance with these notification requirements will not relieve it from making other reports, as required by other State or local laws.

Q6: Since facilities are also broken into in order to obtain equipment or valuables other than radioactive materials, do licensees need to notify the LLEA and the NRC Operations Center every time there is a break-in?

A6: No. The regulation does not require a licensee to request LLEA assistance and to notify NRC offices except in response to an actual or attempted theft, sabotage, or diversion of a category 1 or category 2 quantity of radioactive material. Thus, the licensee's decision on whether to call the police and affected regulators would depend on an assessment of whether the intent of the detected break-in was to perpetrate an unauthorized use of the radioactive material. This assessment of intent could, in turn, depend, among other things, on the proximity of the detected intrusion to any security zone that the licensee has established to isolate the radioactive material from unauthorized access. A licensee's assessment and response to an intrusion alarm in an office supply stockroom could be entirely different from its assessment and response to an intrusion alarm in the radioactive materials storage area. However, the NRC would expect the licensee's security plan and procedures to address ways to distinguish an actual or attempted theft, sabotage, or diversion of the subject radioactive material from a break-in for other purposes. Although the licensee would probably still call the LLEA for a break-in to the office area, it would not need to notify the NRC unless radioactive materials were involved.

Q7: If someone reports lost or missing category 1 or category 2 quantities of radioactive material, how would the NRC respond?

A7: If such an incident occurs, the NRC will expect the licensee to implement the appropriate elements of its security plan. The NRC will also monitor the situation to ensure that the licensee is taking the appropriate actions to locate and recover the missing material. The National Response Framework requires the NRC to notify and coordinate with other Federal agencies as needed.

§ 37.57, "Reporting of Events" (continued)

§ 37.57(b)

The licensee shall assess any suspicious activity related to possible theft, sabotage, or diversion of category 1 or category 2 quantities of radioactive material and notify the LLEA as appropriate. As soon as possible but not later than 4 hours after notifying the LLEA, the licensee shall notify the NRC Operations Center ((301) 816-5100).

Explanation

A licensee must assess any suspicious activity related to the possible theft, sabotage, or diversion of radioactive material and must notify the LLEA as appropriate. The licensee must also notify the NRC Operations Center as soon as possible, but not later than 4 hours after notifying the LLEA.

Q&As

Q1: Why am I required to report suspicious activities to the NRC?

A1: The reporting of suspicious activities is important for evaluating the threat against licensed facilities and material. The NRC reviews individual notifications of suspicious activities to evaluate if potential preoperational activities (i.e., multiple events at a single site or multiple events at multiple sites) may be part of a larger plan and to integrate this information with other agencies in the homeland security and intelligence communities. The NRC is not asking licensees to actively gather intelligence but instead to report information that they believe is relevant to the security of their facility or activity. The reporting requirements provide a consistent means of communicating this information to the NRC.

Q2: Do I need to call the NRC after the LLEA for issues not related to my licensed activities with radioactive materials?

A2: No. The licensee does not need to contact the NRC for such issues unless it concludes that these issues could affect the security or safety of its materials (e.g., if the licensee discovers an unauthorized weapon, drugs, or other contraband inside a controlled area). The NRC broadly worded both the regulations and guidance in this area to provide licensees flexibility.

Q3: Where can I find additional guidance on what the NRC considers suspicious activities that should be reported?

A3: Annex C, "Examples of Reportable Suspicious Activities under 10 CFR 37.57(b)," to Subpart C provides examples of activities that could be considered suspicious activities;

however, these are not all-inclusive examples. Each licensee is responsible for developing and periodically reassessing its own criteria for identifying suspicious activities based on local conditions, recent events, and lessons learned by others. For this purpose, licensees may find the Safeguard New York publication of the New York State Division of Homeland Security and Emergency Services useful because it provides examples of suspicious activities and indicators for heightened surveillance. In addition, a more exhaustive analysis can be found in a 2006 University of Arkansas study, funded by the National Institute of Justice, entitled, "Pre-Incident Indicators of Terrorist Incidents: The Identification of Behavioral, Geographic, and Temporal Patterns of Preparatory Conduct," issued March 2006. Numerous other sources are available on and off the Internet as well.

Q4: If I don't think an unusual activity is "suspicious" and don't report it, and it turns out to be related to a security incident, will I be in violation of this requirement?

A4: No. The NRC recognizes that judgments about what is "suspicious" are inherently subjective and are often influenced by personal experience and community history. After an event, it is more important that the licensee learn from it and take measures to prevent its reoccurrence than it is to assign individual blame. For this reason, the licensee must report, even if after the fact, any unusual but seemly nonthreatening activity that, in retrospect, is found to have been a precursor to an actual or attempted theft, sabotage, or diversion. The NRC, law enforcement agencies, and the intelligence community need this information to help identify the likely perpetrators of similar tactics in the future and to integrate as much credible evidence as possible into a pattern that might indicate an ongoing adversary strategy. Prompt licensee reporting of suspicious activities also enables the NRC to alert other licensees to look for similar activities as possible early indicators of a pending attack.

Q5: What if one of my employees or contractors thinks *everything* is suspicious and frequently reports any nonroutine activity as suspicious? How can I discourage over-reporting that could have the counterproductive effect of making my LLEA(s) *less* inclined to respond?

A5: The licensee can develop criteria for identifying suspicious activities based on Annex C and any other pertinent sources, including one or more of those cited above. The licensee should incorporate these criteria into its site security procedures, should make these procedures available to all facility staff and contractors, and should require all observers of suspicious activities to refer to the procedures in after-action reports. Depending on the size of the licensee's workforce and on the possibility of unwanted chilling effects on reporting, the licensee can also require a first observer to report to the site security officer or his or her designee before reporting to the LLEA and the NRC. In addition, note that 10 CFR 37.57(b) does not require the reporting of all activities that might at first appear suspicious; the regulation requires reporting "as appropriate," which allows licensees to investigate before deciding if the activity warrants reporting.

§ 37.57, "Reporting of Events" (continued)

§ 37.57(c)

The initial telephonic notification required by paragraph (a) of this section must be followed within a period of 30 days by a written report submitted to the NRC by an appropriate method listed in § 37.7.

The report must include sufficient information for NRC analysis and evaluation, including identification of any necessary corrective actions to prevent future instances.

Explanation

The licensee must submit a written report within 30 days to the NRC for its analysis and evaluation. The report must identify any necessary corrective actions.

Q&As

Q1: Other than identifying "any necessary corrective actions," what, at a minimum, should a licensee's written report on a security incident contain in order to provide "sufficient information for NRC analysis and evaluation"?

A1: The content and detail of the report would depend on a number of considerations, including the nature and severity of the security incident, whether it was a first occurrence or a reoccurrence and if it has a significant potential to recur, the adequacy of the licensee's monitoring efforts, the timeliness of the initial detection, the accuracy of the assessment, and the timeliness and potential effectiveness of the measures implemented in response to the incident. If the incident had occurred for the first time, the report should discuss whether the licensee's security plan and procedures had anticipated the incident. In addition to an identification of corrective actions, the report should describe the data and analyses that supported the licensee's identification of the root and significant contributing cause(s) of the incident and should explain how these data and analyses supported the licensee's selection of corrective actions identified. If the incident was a recurrence of a similar incident, the report should briefly describe past corrective actions and the findings of past reassessments and should identify likely reasons that the past corrective actions have not prevented a recurrence of the condition. The report should then explain the basis for the licensee's determination that the proposed new or revised corrective actions or that changes in the licensee's implementation of these actions are likely to prevent a recurrence of the condition or to mitigate its effects.

Annex C. Examples of Reportable Suspicious Activities under 10 CFR 37.57(b)

A licensee should not consider security events reported under 10 CFR 37.57(b) as performance failures; instead, the U.S. Nuclear Regulatory Commission (NRC) considers timely and comprehensive communication of matters relating to threats, attacks, or suspicious activities a vital component of its efforts to assess the current threat environment. Because adversaries of the United States have demonstrated an ability to attack multiple independent targets, timely reporting of nonthreatening, but suspicious, activities is important to the NRC, law enforcement agencies, and the intelligence community in their efforts to integrate potential adversary plans, intentions, and suspicious event reports into a current threat assessment. Prompt reporting of actual or imminent hostile actions permits the NRC to execute its strategic missions of communicating to senior Federal officials and other licensees about hostile actions against the facilities and activities that the agency regulates, thereby protecting public health and safety, the common defense and security, and the environment.

In accordance with 10 CFR 37.57(b), the licensee should report the following security-related events involving suspicious activity that may indicate preoperational surveillance, reconnaissance, or intelligence-gathering activities directed against licensees or their facilities:

- an individual(s) with nonroutine interests or inquiries related to security measures; personnel or vehicle entry points and access controls; or vehicle barrier systems, including fences, walls, or other barriers
- an individual(s) conducting unapproved photographing or videotaping of licensed facilities
- suspicious attempts to recruit or compromise employees or staff and licensee contractors, who are knowledgeable of key personnel, facilities, or systems, into providing safeguards information-modified or other sensitive physical security
- an individual who loiters for no apparent purpose in areas in which he or she could gather intelligence or could perform preoperational reconnaissance
- an individual who displays suspicious behavior (e.g., fleeing, moving quickly away from the licensee or certificate-holder personnel, or unexpected vehicular movement)
- an individual who is secretively sketching, making a map, or taking notes on the facility
- an individual who is eliciting information from security or other site personnel on security systems or vulnerabilities
- unusual challenges to security systems that could represent attempts to gather information on system performance or personnel or equipment response actions
- unauthorized attempts to probe or gain access to the licensee's business secrets or other sensitive information or to control systems, including the use of social engineering techniques (e.g., impersonating authorized users)
- theft or suspicious loss of official company identification documents, uniforms, or vehicles necessary to access plant facilities
- use of forged, stolen, or fabricated documents to support access control or authorization activities
- unusual attempts to obtain information or documents related to site security training, techniques, procedures, or practices
- discovery of Internet site postings that make violent threats related to specific licensed facilities or activities
- unusual threats or terrorist-related activities that become known to facility security or management involving (1) unusual surveillance, probing, or reconnaissance, (2) attempts to gain unauthorized access, (3) attempts to gain access to, or acquire, hazardous or dangerous materials, unusual use of materials, or (5) financing to support terrorist activities
- stated threats against the licensee's facility or staff, unless the licensee determines that the threats are unsubstantiated

- unsubstantiated bomb or extortion threats that are considered to be related to harassment, including those that represent tests of response capabilities or intelligence-gathering activities or an attempt to disrupt facility operations
- fires or explosions of suspicious or unknown origin
- the unauthorized operation, manipulation, or tampering of radioactive material in quantities of concern or the unauthorized operation, manipulation, or tampering of security-related structures, systems, and components that could prevent the implementation of the licensee's protective strategy
- the intentional cutting of wires that does not affect the facility or security operations
- the modification of security equipment that renders the equipment inoperable
- the overt changing of equipment or control settings

The NRC does not consider this list to be exclusive or exhaustive. Additionally, licensees should evaluate an event that is not reportable under this requirement for reporting or recording under the other provisions of 10 CFR Part 37.

Subpart D—Physical Protection in Transit

§ 37.71, "Additional Requirements for Transfer of Category 1 and Category 2 Quantities of Radioactive Material"
§ 37.73, "Applicability of Physical Protection of Category 1 and Category 2 Quantities of Radioactive Material during Transit"
§ 37.75, "Preplanning and Coordination of Shipment of Category 1 or Category 2 Quantities of Radioactive Material"
§ 37.77, "Advance Notification of Shipment of Category 1 Quantities of Radioactive Material"
§ 37.79, "Requirements for Physical Protection of Category 1 and Category 2 Quantities of Radioactive Material during Shipment"
§ 37.81, "Reporting of Events"

§ 37.71, "Additional Requirements for Transfer of Category 1 and Category 2 Quantities of Radioactive Material"

A licensee transferring a category 1 or category 2 quantity of radioactive material to a licensee of the Commission or an Agreement State shall meet the license verification provisions listed below instead of those listed in 10 CFR 30.41(d) of this chapter:

§ 37.71(a)

Any licensee transferring category 1 quantities of radioactive material to a licensee of the Commission or an Agreement State, prior to conducting such transfer, shall verify with the NRC's license verification system or the license issuing authority that the transferee's license authorizes the receipt of the type, form, and quantity of radioactive material to be transferred and that the licensee is authorized to receive radioactive material at the location requested for delivery. If the verification is conducted by contacting the license issuing authority, the transferor shall document the verification.

For transfers within the same organization, the licensee does not need to verify the transfer.

§ 37.71(b)

Any licensee transferring category 2 quantities of radioactive material to a licensee of the Commission or an Agreement State, prior to conducting such transfer, shall verify with the NRC's license verification system or the license issuing authority that the transferee's license authorizes the receipt of the type, form, and quantity of radioactive material to be transferred. If the verification is conducted by contacting the license issuing authority, the transferor shall document the verification. For transfers within the same organization, the licensee does not need to verify the transfer.

§ 37.71(c)

In an emergency where the licensee cannot reach the license issuing authority and the license verification system is nonfunctional, the licensee may accept a written certification by the transferee that it is authorized by license to receive the type, form, and quantity of radioactive material to be transferred. The certification must include the license number, current revision number, issuing agency, expiration date, and, for a category 1 shipment, the authorized address. The licensee shall keep a copy of the certification. The certification must be confirmed by use of the NRC's license verification system or by contacting the license-issuing authority by the end of the next business day.

§ 37.71(d)

The transferor shall keep a copy of the verification documentation as a record for 3 years.

Explanation

Before transferring category 1 or category 2 quantities of radioactive material, the shipping licensee must verify that the transferee's license is valid. The licensee can conduct this verification by using the NRC's license verification system (LVS) or by contacting the agency (the NRC or Agreement State) that issued the transferee's license. The licensee must maintain verification documentation for 3 years.

Q&As

Q1: Is verification of the transferee's license necessary?

A1: Yes. The regulations in 10 CFR 37.71 require the licensee to verify that recipients of category 1 or category 2 quantities of radioactive material are licensed and are authorized to receive them. The regulation in 10 CFR 37.71(a) requires the verification to occur before the radioactive material is actually transferred. Verification is not required for transfers within the same organization or company.

Q2: What information needs to be verified?

A2: The regulation in 10 CFR 37.71(a) requires the licensee to verify that the transferee's license authorizes receipt of the type, form, and quantity of radioactive material that will be transferred. For transfers of category 1 quantities of radioactive material, the shipping licensee must also verify that the receiving licensee is authorized to receive radioactive material at the address requested for delivery.

Q3: How does a licensee conduct the verification?

A3: The licensee could conduct these verifications by using the NRC's LVS or, if the system is not available, by contacting the license-issuing authority (i.e., the NRC or the appropriate Agreement State agency). Licensees should contact the appropriate NRC regional office to verify the validity of NRC licensees. Information on Agreement State contacts appears on the NRC Web page at http://nrc.stp.ornl.gov/asdirectory.html. Use of the LVS when it becomes available will be the preferred method for license verification.

Q4: May I use a fax or e-mail to verify the validity of a license, or just keep a copy of the recipient's license on file?

A4: No. Neither facsimile or e-mail provide the positive identification needed for verifying the validity of a license because an individual can alter an electronic image of the license to change its possession limits, authorized location of use, or even the name of the person who received the license. Keeping a copy of the recipient's license on file also is inadequate for verifying the validity of a license because the license could be amended or terminated. The licensee shipping the material would not know that the intended recipient's license was no longer valid. The regulation in 10 CFR 37.71(a) requires the licensee to use the LVS or to contact the regulatory agency (the NRC or Agreement State) to verify that a license is valid before shipping category 1 or category 2 quantities of radioactive material to a domestic company.

Q5: What is the License Verification System?

A5: The LVS is a new Web-based NRC system designed to enable users to verify electronically the validity of a license issued by either the NRC or an Agreement State. Use of the LVS is the preferred method for license verification. If the system is not available, licensees would need to contact the appropriate licensing agency. More information will be available once the system is fully developed and operational.

Q6: If the licensee is transferring the material to DOE or to an entity that does not have an NRC or Agreement State license (e.g., for exporting the material), does it still need to verify the transfer?

A6: No. The license verification requirement is only applicable if the licensee is transferring category 1 or category 2 quantities of radioactive material to a licensee of the NRC or an Agreement State. License verification is not required to transfer material to DOE or to any other Federal entity that does not have an NRC or Agreement State license. However, licensees exporting material would need to meet the requirements in 10 CFR Part 110, "Export and Import of Nuclear Equipment and Material," for checking the documentation that the recipient has the necessary authorization under the laws and regulations of the importing country.

Q7: Why are there differences between the category 1 and category 2 verification requirements?

A7: Because category 1 quantities of radioactive material are of greater concern than category 2, the regulation in 10 CFR 37.71(a) subjects the licensee that is transferring category 1 quantities to the additional requirement of verifying that the receiving licensee is authorized to receive radioactive material at the address requested for delivery.

Q8: When should a licensee document the verification, and how should it be documented?

A8: The rule only requires documentation if the shipping licensee (the licensee sending the licensed material) conducts the verification by contacting the NRC or Agreement State license-issuing authority. If the licensee uses the LVS, the system will save the record; the licensee would not need to keep further documentation. If the licensee contacts the license-issuing authority, it should document the communication. If the contact was by phone, the licensee should prepare a note to file that contains information on the transfer (i.e., the company name, the license number, the material being transferred, and the address for category 1 shipments), the date of the communication, and the name of the individual at the licensing agency. If conduct of the communication was done by e-mail or facsimile, the licensee should retain a copy of the e-mail or facsimile from the licensing agency.

Q9: If I can't reach the regulatory agency or access the LVS to verify that the recipient of my shipment is licensed to receive and possess the material, do I have to postpone the shipment until I obtain the verification?

A9: No. The rule provides an alternative that allows licensees to ship in situations in which the regulator cannot be reached or the LVS is inaccessible. Under 10 CFR 37.71(c), the shipping licensee may obtain a written certification from the receiving licensee noting that it is authorized to receive the requested radioactive material. The certification must include the license number; current revision number; issuing agency; expiration date; and, for a category 1 shipment, the authorized address. The shipping licensee must follow up with LVS or the license-issuing authority by the end of the next business day to confirm that the recipient's license is valid. A copy of the certification must be retained. The licensee can fax or e-mail the certification, and receipt of the certification through the mail is not necessary.

Q10: If I rely on the recipient licensee's certification to ship material under the provisions of 10 CFR 37.71(c) for emergencies, what do I do if I discover that the recipient's license is not valid for the transfer after the shipment has been sent?

A10: First, the licensee should contact the regulatory agency to determine if the recipient's license authorizes the shipment. If the licensee used the LVS, the system may not have yet updated the information to reflect a license amendment. If the regulatory agency confirms that the recipient is not authorized to receive the material, the licensee should discuss, with the regulatory agency, what actions should be taken. One possible action is to contact the carrier to determine if it has delivered the shipment. If it has not, the licensee may be able to request that the carrier return the shipment. If the shipment has been delivered, the licensee may need to notify the LLEA at the licensee's location and the NRC's Operation Center. Another possible action is for the recipient to request a license amendment that would authorize the radioactive material in question.

Q11: Do I need to verify a transfer if I am sending the material to another location within the company?

A11: No. For transfers within the same organization, the licensee does not need to verify the transfer.

Q12: What should I do if I receive an unusual order?

A12: The regulations in 10 CFR Part 37 do not require a shipping licensee to notify its regulatory agency of an unusual order. The licensee should, however, as good practice, carefully review orders from unfamiliar entities, even if their licenses and possession limits can be verified. Similarly, the licensee should, as good practice, carefully review an order from a previous recipient licensee if the order appears unusual compared to previous shipments to that licensee in regard to the type, form, destination, and quantity of material that will be shipped. Unusual orders raise heightened security concerns, including attempts to obtain radioactive materials, and shipping licensees should take appropriate precautions. These precautions include contacting the receiving licensee or the governing regulatory authority or both.

Q13: Typical source or device manufacturer and distributor licensees could ship 50 to 60 sources a day to user licensees in many different jurisdictions. An M&D licensee could thus be making 50 or 60 calls a day—or more, if the M&D licensee hasn't gotten an immediate response on its first call—to get the regulatory authority's verification for each shipment. The same problem could arise for user licensees returning these sources or devices to the M&D licensee; every single user licensee would have to call the appropriate Agreement State or NRC office to verify that the M&D licensee is authorized to receive the sources. Is verification necessary for each shipment, even for returns to the M&D licensee?

A13: Yes. Verification of each shipment is necessary. One of the recommended actions of the 2006 "Radiation Source Protection and Security Task Force Report" was that the NRC consider imposing "additional measures to verify the validity of licenses, before transfer of risk-significant radioactive sources, on all licensees authorized to possess category 1 and 2 quantities of radioactive material" (Action 4-1, page xxvii, pages 48–50). The NRC agrees with the recommended action and maintains that verification of the intended recipient's license before transfer is important to enhance the security of the material by validating the licensee's legitimacy. Using the LVS instead of calling the regulatory agency should significantly reduce the burden and should allow 100-percent validation of licenses before the transfer of category 1 or category 2 quantities of radioactive material. The Radiation Source Protection and Security Task Force is an interagency task force mandated by the EPAct.

§ 37.73, "Applicability of Physical Protection of Category 1 and Category 2 Quantities of Radioactive Material during Transit"

§ 37.73(a)

For shipments of category 1 quantities of radioactive material, each shipping licensee shall comply with the requirements for physical protection contained in §§ 37.75(a) and (e); 37.77; 37.79(a)(1), (b)(1), and (c); and 37.81(a), (c), (e), (g) and (h).

§ 37.73(b)

For shipments of category 2 quantities of radioactive material, each shipping licensee shall comply with the requirements for physical protection contained in §§ 37.75(b) through (e); 37.79(a)(2), (a)(3), (b)(2), and (c); and 37.81(b), (d), (f), (g), and (h). For those shipments of category 2 quantities of radioactive material that meet the criteria of § 71.97(b) of this chapter, the shipping licensee shall also comply with the advance notification provisions of § 71.97 of this chapter.

§ 37.73(c)

The shipping licensee shall be responsible for meeting the requirements of this subpart unless the receiving licensee has agreed in writing to arrange for the in-transit physical protection required under this subpart.

§ 37.73(d)

Each licensee that imports or exports category 1 quantities of radioactive material shall comply with the requirements for physical protection contained in §§ 37.75(a)(2) and (e); 37.77; 37.79(a)(1), (b)(1), and (c); and 37.81(a), (c), (e), (g), and (h) during the domestic portion of the shipment.

§ 37.73(e)

Each licensee that imports or exports category 2 quantities of radioactive material shall comply with the requirements for physical protection during transit contained in §§ 37.79(a)(2), (a)(3), (b)(2), and (d); and 37.81(b), (d), (f), (g), and (h) during the domestic portion of the shipment.

Explanation

This section establishes which provisions apply for shipments of category 1 and category 2 quantities of radioactive material. The shipping licensee is responsible for meeting the requirements unless the receiving licensee agrees in writing to assume responsibility. For ease of reference, see the table below.

Applicability Table for Subpart D Requirements

	§ 37.75(a)	§ 37.75(a)(2)	§ 37.75(b)	§ 37.75(c)	§ 37.75(d)	§ 37.75(e)	§ 37.77	§ 37.79(a)(1)	§ 37.79(a)(2)	§ 37.79(a)(3)	§ 37.79(b)(1)	§ 37.79(b)(2)	§ 37.79(c)	§ 37.81(a)	§ 37.81(b)	§ 37.81(c)	§ 37.81(d)	§ 37.81(e)	§ 37.81(f)	§ 37.81(g)	§ 37.81(h)
Category 1 shipments	x	x				x	x	x			x		X	x		x		x		x	x
Category 2 shipments*			x	x	x	x			x	x		x	X		x		x		x	x	x
Domestic portion of Category 1 imports/ exports		x				x	x	x			x		X	x		x		x		x	x
Domestic portion of Category 2 imports/ exports									x	x		x			x		x		x	x	x

*For shipments of category 2 quantities that meet the criteria of 10 CFR 71.97(b), the shipping licensee must also comply with the advance notification provisions of 10 CFR 71.97, "Advanced Notification of Shipment of Irradiated Reactor Fuel and Nuclear Waste."

Q&As

Q1: Would Subpart D requirements apply to a shipment of two or more packages if each package contains less than a category 2 quantity but in aggregate equals or exceeds a category 2 quantity?

A1: Yes. Each NRC licensee must meet the requirements in Subpart D for any radioactive material quantity that meets or exceeds a category 2 quantity regardless of how many individual packages may be in the shipment. This question assumes that the material in the two or more individual packages aggregates to a category 2 quantity, and the regulation in 10 CFR 37.73(c) provides that "[f]or shipments of category 2 quantities of radioactive material, each shipping licensee shall comply" with the requirements in Subpart D referenced in this subsection. Because this provision applies to all shipments and does not refer to the number of packages in each shipment, a licensee cannot avoid meeting these requirements by partitioning the total shipping quantity into multiple packages on the same vehicle. An adversary (including an insider) in control of the transport vehicle would have unauthorized access to the total quantity of radioactive material in the shipment.

Q2: Which licensee is responsible for meeting the requirements of Subpart D?

A2: Generally, the shipping licensee is responsible for meeting the requirements in Subpart D. However, the receiving licensee may choose to agree, in writing, to arrange for the required in-transit physical protection measures. In that case, the licensee that receives the shipment would be responsible for compliance with the physical protection requirements of this subpart.

Q3: During what portion of the shipment are the security requirements applicable for category 1 or category 2 quantities of imports?

A3: The category 1 and category 2 import requirements are applicable only from the point that the material enters the United States (i.e., the domestic portion of the shipment after the package clears U.S. Customs and Border Protection).

Q4: For material being exported, is the licensee required to follow the security requirements for the entire trip?

A4: No. The licensee is responsible only for following the security provisions for the domestic portion of the trip until the shipment comes under the jurisdiction of a U.S. Government agency (e.g., the Federal Aviation Administration or DHS) at a port, border crossing, or airport.

Q5: Do the security provisions for category 1 and category 2 quantities of radioactive material apply during the time when shipments are placed in interim storage at the shipping licensee's facility awaiting pickup?

A5: Yes. The security provisions do apply. Licensees that ship category 1 or category 2 quantities should ensure that any interim storage is minimized or, when it cannot be avoided, that required security provisions are met.

Q6: Do the security provisions apply to transshipments?

A6: No. The security provisions of Subpart D do not apply to transshipments of category 1 or category 2 quantities of radioactive material. Transshipments are shipments that are originated by a foreign entity in one country, pass through the United States, and continue on to a recipient in another country.

§ 37.75, "Preplanning and Coordination of Shipment of Category 1 or Category 2 Quantities of Radioactive Material"

§ 37.75(a)

Each licensee that plans to transport, or deliver to a carrier for transport, licensed material that is a category 1 quantity of radioactive material outside the confines of the licensee's facility or other place of use or storage shall:

§ 37.75(a)(1)

Preplan and coordinate shipment arrival and departure times with the receiving licensee;

§ 37.75(a)(2)

Preplan and coordinate shipment information with the governor or the governor's designee of any State through which the shipment will pass to:

§ 37.75(a)(2)(i)

Discuss the State's intention to provide law enforcement escorts; and

§ 37.75(a)(2)(ii)

Identify safe havens; and

§ 37.75(a)(3)

Document the preplanning and coordination activities.

Explanation

Each licensee that ships category 1 quantities of radioactive material must conduct preplanning and coordination activities with the receiving licensee and with each State that the shipment crosses.

Q&As

Q1: What happens if a shipment has to be rerouted during transport because of bad weather or other developments that neither I nor my carrier can control? How can I be sure in advance of the States that my category 1 shipment will pass through and that I'll need to coordinate with under this subsection?

A1: The licensee can be reasonably sure about the alternative routes for a shipment because the carrier is subject to Federal DOT regulations under 49 CFR 97.101, "Requirements for Motor Carriers and Drivers," on highway-route-controlled quantities of radioactive materials. These regulations require carriers to minimize the shipment's time in transit and to use a preferred route, which is defined as "an Interstate System highway," unless a State-routing agency designates an alternative route. To ensure its compliance with

10 CFR 37.75(a) requirements for the preplanning and coordination with States through which the shipment will pass, the licensee should, as good practice, ensure that its shipment contract with the carrier clearly obligates the carrier to also comply with these requirements when an unusual event, such as a blizzard, requires the rerouting of the shipment through a State other than those along the originally planned route.

Q2: What type of documentation is required under 10 CFR 37.75(a)(3)?

A2: The regulation in 10 CFR 37.75(a)(3) requires licensee to document the preplanning and coordination activities required by 10 CFR 37.75(a)(1) and 10 CFR 37.75(a)(2) for shipments of category 1 quantities of radioactive materials. The shipping licensee should therefore document any phone conversations or e-mail communications that it has with the receiving licensee to include the names of the individuals participating in the call or e-mail communications, a general description of the shipment, and the departure and arrival times. The shipping licensee should also document any interactions with the governor's designee to include the names of the individuals participating in the call or e-mail, the route-affected States' decisions on escorts, the safe havens identified, and any other information that the licensee considers pertinent to document compliance with the requirements in this subsection.

Q3: What does the NRC consider to be a safe haven?

A3: A safe haven is defined as "[a] readily recognizable and readily accessible site at which security is present or from which, in the event of an emergency, the transport crew can notify and wait for the LLEAs." For additional discussion of this definition, see the pertinent Q&As under 10 CFR 37.5.

Licensees should use the following criteria to identify safe havens for shipments:

- The safe haven is near the route (i.e., readily available to the transport vehicle).
- Security from local, State, or Federal assets is present or is accessible for a timely response.
- The site is well lit, has adequate parking, and can be used for emergency repair or for waiting for the LLEA response on a 24-hour basis.
- Additional telephone facilities are available if the communications system of the transport vehicle fails to function properly.
- Possible safe haven sites include Federal sites that have significant security assets, such as military-base gates or guarded agency parking lots; secure company terminals; State weigh stations; State welcome stations or rest areas; scenic overlooks or visitors' centers; truck stops with secure areas; and LLEA sites, including State police barracks.

Identifying safe havens is the licensee's responsibility. If a licensee is having difficulty identifying a safe haven, it can contact the NRC, the appropriate Agreement State agency, or State or local law enforcement officials.

Q4: Must I have more than one safe haven for a category 1 quantity shipment? How many should I identify?

A4: The rule does not specify a minimum number of safe havens for any shipment, but the shipping licensee and carrier have an important interest in identifying available places to stop along the route on which the transport vehicle could securely remain or safely wait for assistance.

Q5: How far apart should these safe havens be? Is there a minimum distance for which at least one safe haven is required?

A5: The rule does not specify a minimum distance between safe havens. The NRC recommends 50 miles. However, the agency recognizes that safe havens at that interval may not be available in some more remote areas. In most areas, commercial truck stops, service stations, State weigh stations, welcome stations, rest areas, and paved scenic overlooks or visitor parking for historic sites should provide ample alternatives to shipment preplanners and coordinators.

Q6: Is there a list I can refer to of approved safe havens?

A6: No. However, if a licensee is having difficulty identifying a safe haven, it may contact the NRC, the appropriate Agreement State agency, or State or local law enforcement officials.

Q7: If a State imposes other transportation security requirements outside the scope of 10 CFR Part 37, do I have to meet those, too?

A7: Yes. Unless any DOT requirement under Chapter 49, "Transportation," of the *Code of Federal Regulations* preempts those security requirements, any State may apply additional requirements beyond those necessary for Agreement State compatibility under 10 CFR Part 37 while the shipment is within the State's borders. The regulations in 10 CFR Part 37 do not preempt other State requirements that may be applicable, such as State police escorts for radioactive material shipments. Licensees must comply with these State requirements and the requirements of the NRC or the Agreement State agency with jurisdiction.

§ 37.75, "Preplanning and Coordination of Shipment of Category 1 or Category 2 Quantities of Radioactive Material" (continued)

§ 37.75(b)

Each licensee that plans to transport, or deliver to a carrier for transport, licensed material that is a category 2 quantity of radioactive material outside the confines of the licensee's facility or other place of use or storage shall coordinate the shipment no-later-than arrival time and the expected shipment arrival with the receiving licensee. The licensee shall document the coordination activities.

Explanation

Licensees shipping category 2 quantities of radioactive material must coordinate the expected arrival time and the NLT arrival time with the receiving licensee.

Q&As

Q1: If I'm a shipping licensee, how soon after I "plan to transport" must I coordinate with a receiving licensee?

A1: The licensee may begin its coordination activities with the receiving licensee at any time after it has decided to ship the material. However, under this subsection, the licensee must complete these activities before it can transport the material or before it consigns the material to a carrier for transport. The licensee does not need to begin coordinating with the receiving licensee as soon as it confirms an incoming order or decides to transfer the material.

Q2: What does it mean to "coordinate," and how much is enough?

A2: Under this subsection, the licensee only needs to contact the receiving licensee and to establish an expected arrival time and NLT arrival time for the shipment. This coordination can be as simple as sending an e-mail stating that the shipment is expected to arrive via FedEx by noon, for example, on a specified date and that the NLT arrival time is 6 p.m. on that day. Although this coordination is the only kind required, the licensee and the recipient licensee are free to negotiate any other understandings, such as times and methods of communication; contact information for the carrier; contact information for LLEAs through which the shipment will pass; or any other information that both parties agree would better enable them to anticipate, and to be prepared for, sources of delay or would otherwise better enable them to make the shipment as incident-free as possible. Coordination may be done by phone, facsimile, e-mail, or face-to-face meetings; however, the licensee must document this activity and the agreements that it produces, as required by this subsection.

Q3: We have two facilities and need to ship a category 2 quantity of radioactive material from one facility across town to the other. Do we still need to "coordinate" an estimated and a NLT arrival time, even though we're coordinating with ourselves?

A3: If the licensee's shipping and receiving facilities operate under different licenses and if the licensee is using a commercial carrier to transport the material, coordination must be done like it is in all cases involving different licensees. If the licensee's transferring and receiving facilities are operating under the same license and if the licensee is providing its own transportation, coordination with the receiving facility is not required. However, the licensee may want to consider some elements of coordination, such as having the shipping facility notify the destination facility of the shipment's departure and its estimated arrival time, if it believes that such coordination might be prudent because, for example, the two facilities are separated by property or roads that are not under its control.

Q4: We are part of a university and plan to ship a category 2 quantity of radioactive material from one facility to another on the same campus. Do we still need to coordinate an estimated and NLT arrival time between the two onsite facilities? What if the two facilities operate under different licenses?

A4: Facilities on the same site that operate under the same license do not need to coordinate even if the facilities are not at contiguous locations and even if the material has to cross or travel on a public road that crosses the site. However, if the shipping and receiving facilities are different licensees, they must coordinate the estimated and NLT arrival time because possession of the material will be transferred from one licensee to another.

Q5: Does an industrial radiographer carrying a camera containing a category 2 quantity sealed source have to coordinate under this section to transport the source from a company site to a temporary jobsite or from one temporary jobsite to another?

A5: A licensed industrial radiographer is not subject to these coordination requirements for either kind of shipment because the radiographer is required to maintain control of a mobile source at all times, and transfer of the source from one licensee to another at the company site or at either of the temporary jobsites does not occur.

Q6: What is meant by NLT arrival time?

A6: As defined in 10 CFR 37.5 of this rule, NLT arrival time means "the date and time that the shipping licensee and receiving licensee have established as the time at which an investigation will be initiated if the shipment has not arrived at the receiving facility." Licensees do not have to begin an investigation if the shipment does not arrive by the estimated arrival time. For a category 2 shipment, the NLT arrival time may not be more than 6 hours after the estimated arrival time. The NLT date and time effectively requires verification that the shipment arrived within the expected timeframe. If the shipment has not arrived by the agreed upon NLT arrival time, the shipping licensee must trace the location of the shipment and must decide whether to report to affected law enforcement authorities an unusual occurrence that could lead to a theft or diversion of the material.

Q7: What types of records are required to document coordination activities?

A7: The last sentence of this subsection requires a licensee to "document the coordination activities" required to coordinate the shipment NLT arrival time and the expected shipment arrival with the receiving licensee. The shipping licensee should document any phone conversations or e-mail correspondence with the receiving licensee, including the names of the individuals participating in the calls or correspondence and the agreed upon estimated arrival time and NLT arrival time. The licensee may print or store e-mails electronically, and it should maintain a copy of any facsimile and the confirmation information.

Q8: Am I allowed to use the NSTS as a method to fulfill the preplanning and coordination requirements of 10 CFR 37.75(b)?

A8: No. Although the National Source Tracking System (NSTS) is used, among other things, to track transfers of radioactive materials that are subject to the reporting requirements in Appendix E, "Nationally Tracked Source Thresholds," to 10 CFR Part 20, the NSTS is not designed for the kind of detailed interactions that are necessary for effective preshipment planning and coordination. For example, the NSTS does not track estimated or NLT arrival times; it only tracks the dates of estimated and actual arrivals. Reporting is not required until after the transfer has been made. This deadline alone could make the NSTS useless for determining when and whether to begin tracing a missing shipment. In addition, the NSTS does not require the reporting of shipments that include bulk material or that comprise several category 3 quantities that add up to a category 2 quantity. Thus, the NSTS does not cover all shipments that are subject to 10 CFR Part 37 requirements.

Q9: I'm an NRC licensee shipping a category 2 quantity of material to an Agreement State licensee, and that State's regulatory agency hasn't yet adopted the new regulations required to implement 10 CFR Part 37. 10 CFR 37.75(c) requires the receiving Agreement State licensee to notify me if the shipment hasn't arrived by the NLT arrival time, and I'm required under 10 CFR 37.79(c) to investigate "immediately" if the shipment has missed that deadline. But if the receiving licensee is operating under existing Agreement State regulations, it isn't required to notify me when a shipment misses its NLT arrival time. Will I be in violation of my NRC license if I don't start investigating by that time?

A9: This should not be an issue. The security orders issued by the NRC and the legally binding requirements issued by the Agreement States require licensees to confirm receipt of transferred radioactive material; if the recipient licensee does not receive the material at the expected time of delivery, it must notify the originator and assist in any investigation. This notification would actually occur earlier than the notification required by 10 CFR Part 37.

§ 37.75, "Preplanning and Coordination of Shipment of Category 1 or Category 2 Quantities of Radioactive Material" (continued)

§ 37.75(c)

Each licensee who receives a shipment of a category 2 quantity of radioactive material shall confirm receipt of the shipment with the originator. If the shipment has not arrived by the no-later-than arrival time, the receiving licensee shall notify the originator.

Explanation

The receiving licensee must confirm receipt of a shipment with the shipping licensee. The receiving licensee must also notify the shipping licensee if the shipment has not arrived by the NLT arrival time.

Q&As

Q1: When must the receiving licensee confirm to the shipping licensee that the shipment has arrived?

A1: For a category 2 quantity, the receiving licensee should strive to notify the shipping licensee as soon as practicable after the shipment has arrived. It must, however, report that the shipment has not arrived by the agreed upon NLT arrival time. The definition of NLT arrival time in 10 CFR 37.5 states that this interval may not be more than 6 hours after the estimated arrival time. If the shipment arrives on time, confirmation by the delivery service (e.g., FedEx, DHS, or UPS) may be acceptable if the shipping and receiving licensees have agreed to this method during their preplanning and coordination discussions.

Q2: How should the receiving licensee notify the shipping licensee?

A2: The regulations in 10 CFR Part 37 do not specify a particular way for the receiving licensee to notify the shipping licensee. The receiving licensee may contact the shipping licensee by phone, e-mail, or facsimile. The licensees should decide on the method of notification that they will use during their preplanning and coordination activities.

Q3: Am I allowed to use the NSTS as a method to fulfill the requirements of 10 CFR 37.75(c) to confirm receipt of the category 2 shipment and notify the shipping licensee if the shipment has not arrived by its NLT arrival time?

A3: No. The use of NSTS will neither meet the requirement in 10 CFR 37.75(c) to notify the shipping licensee of the receipt of a shipment nor meet the requirement to report to NSTS. A report to the NSTS is a report to a system; it does not notify the shipping licensee of the receipt of a source.

§ 37.75, "Preplanning and Coordination of Shipment of Category 1 or Category 2 Quantities of Radioactive Material" (continued)

§ 37.75(d)

Each licensee, who transports or plans to transport a shipment of a category 2 quantity of radioactive material, and determines that the shipment will arrive after the no-later-than arrival time provided pursuant to paragraph (b) of this section, shall promptly notify the receiving licensee of the new no-later-than arrival time.

Explanation

If a shipment is delayed, the shipping licensee must notify the receiving licensee promptly of any new NLT arrival time.

Q&As

Q1: How "promptly" must a category 2 shipping licensee notify the receiving licensee if the NLT arrival time must be changed?

A1: The shipping licensee's notification should be as soon as practicable to avert a needless alarm by the receiving licensee under 10 CFR 37.75(c) when the shipment is not delivered by the originally established NLT arrival time. More specifically, notification should be as soon as practicable after the driver or other authorized member of the transfer crew determines that the category 2 shipment in question cannot be safely expedited enough to arrive before the NLT arrival time. (The requirement for an NLT arrival time does not apply to shipments of category 1 quantities of radioactive material.)

§ 37.75, "Preplanning and Coordination of Shipment of Category 1 or Category 2 Quantities of Radioactive Material" (continued)

§ 37.75(e)

The licensee shall retain a copy of the documentation for preplanning and coordination and any revision thereof as a record for 3 years.

Explanation

This definition is self-explanatory.

Q&As

Q1: What kinds of preplanning and coordination documents must a licensee retain to satisfy this recordkeeping requirement?

A1: For shipments of category 1 quantities of radioactive materials, the regulation in 10 CFR 37.75(a)(3) requires licensees to document the preplanning and coordination activities required in 10 CFR 37.75(a)(1) and 10 CFR 37.75(a)(2). The shipping licensee should therefore document any phone conversations or e-mail communications that it has with the receiving licensee, including the names of the individuals participating in the call or e-mail communications, a general description of the shipment, and the departure and arrival times. In addition, the shipping licensee should document any interactions that it has with the governor's designee, the names of the individuals participating in the call or e-mail, route-affected States' decisions on escorts, safe havens identified, and any other information that the licensee considers pertinent to document its compliance with the requirements in this subsection. (See also Q3 and A3 for 10 CFR 37.75(a) above.) To document preplanning and coordination activities for shipments of category 2 quantities of radioactive material, the shipping licensee only needs to retain records of the expected arrival time and NLT arrival time. The record should contain the receiving company's name and general information concerning the shipment. (See also Q7 and A7 for 10 CFR 37.75(b) above.)

§ 37.77, "Advance Notification of Shipment of Category 1 Quantities of Radioactive Material"

§ 37.77

As specified in paragraphs (a) and (b) of this section, each licensee shall provide advance notification to the NRC and the governor of a State, or the governor's designee, of the shipment of licensed material in a category 1 quantity, through or across the boundary of the State, before the transport, or delivery to a carrier for transport of the licensed material outside the confines of the licensee's facility or other place of use or storage.

Explanation

Licensees must provide advance notification to the NRC and the governor of any State (or the governor's designee) of a shipment of a category 1 quantity of licensed material that passes through or across the boundary of the State.

Q&As

Q1: What constitutes an "advance notification" for the purposes of this requirement?

A1: An advance notification is a communication to the NRC and the governor (or governor's designated official) of an affected State that a shipment of a category 1 quantity of radioactive material will be made into or within that State's jurisdiction on a set date at a best estimated time. The communication must be written, must follow the procedures in 10 CFR 37.77(a), and must contain the information specified in 10 CFR 37.77(b). The licensee may send the communication by mail, e-mail, or facsimile. Annex D, "Template for Advance Notifications of Shipments to the NRC of category 1 Quantities of Radioactive Material under 10 CFR 37.77(b)," to this subpart provides an example of an advance notification template that would satisfy the NRC.

Q2: Are advance notifications required for any category 2 shipments?

A2: No. Advance notifications are not required for shipments of category 2 quantities of radioactive material unless the shipment falls within the scope of 10 CFR 71.97(b) or unless it is an export or import shipment that requires notification under 10 CFR 110.50(c).

Q3: Where does a licensee obtain contact information for the governor's designee?

A3: A list of the contact information for the governor's designees is published annually, typically in July or August, in the *Federal Register*. The 2011 update appeared on October 31, 2011. An updated list is posted on the NRC Web site at http://nrc-stp.ornl.gov/special/designee.pdf. Copies may also be obtained by contacting the Director, Division of Intergovernmental Liaison and Rulemaking, Office of Federal and State Materials and Environmental Management Programs, U.S. Nuclear Regulatory Commission, Washington, DC 20555. However, note that the currently available list provides governor-designated contact information only for spent nuclear fuel shipments, and the contacts for 10 CFR Part 37 shipments may not be the same. The NRC will add contact information for 10 CFR Part 37 shipments to the latest available listing before the effective date of the rule.

Q4: Whom must the shipping licensee notify?

A4: If the NRC issued the license, the licensee must notify the agency and each State through which the shipment will pass, including the State of the destination. If an Agreement State agency issued the license, the licensee must notify that agency and each State through which the shipment will pass, including the destination State.

Q5: Does the NRC share advance notification information about a shipment with other agencies?

A5: Yes. When the NRC receives advance notification information, it may share the information with other affected Government agencies, such as DOT and DHS.

§ 37.77, "Advance Notification of Shipment of Category 1 Quantities of Radioactive Material" (continued)

§ 37.77(a), "Procedures for Submitting Advance Notification"

§ 37.77(a)(1)

The notification must be made to the NRC and to the office of each appropriate governor or governor's designee. The contact information, including telephone and mailing addresses, of governors and governors' designees is available on the NRC Web site at http://nrcstp.ornl.gov/special/designee.pdf. A list of the contact information is also available upon request from the Director, Division of Intergovernmental Liaison and Rulemaking, Office of Federal and State Materials and Environmental Management Programs, U.S. Nuclear Regulatory Commission, Washington, DC 20555. Notifications to the NRC must be to the NRC's Director, Division of Security Policy, Office of Nuclear Security and Incident Response, U.S. Nuclear Regulatory Commission, Washington, DC 20555. The notification to the NRC may be made by e-mail to RAMQC SHIPMENTS@nrc.gov or by fax to 301-816-5151.

§ 37.77(a)(2)

A notification delivered by mail must be postmarked at least 7 days before transport of the shipment commences at the shipping facility.

§ 37.77(a)(3)

A notification delivered by any means other than mail must reach the NRC at least 4 days before the transport of the shipment commences and must reach the office of the governor or the governor's designee at least 4 days before transport of a shipment within or through the State.

Explanation

Advance notifications of a shipment must be made in writing and postmarked 7 days before the shipment begins. Notifications delivered by any other means than mail must reach the office of the governor or the governor's designee at least 4 days before transport within or through the State.

Q&As

Q1: Why does the mail notification requirement differ from that of other notifications?

A1: Mail delivery is not instantaneous; additional time is required to allow the mail notification to reach the appropriate party by the date required for other means of delivery. Under 10 CFR 37.77(a)(3), this delivery date is "at least 4 days before the transport of the shipment commences" for a notification to the NRC and "at least 4 days before transport of a shipment within or through the State" for a notification to a governor or governor's designee. To make the arrival of the notification by this date more likely to occur, the licensee must send a mailed notification sufficiently early—postmarked at least 7 days before the shipment commences—to allow the recipient to receive it when he or she would have received it if it was sent by other means, such as by facsimile or e-mail.

Q2: Does a licensee have to mail the advance notification via the U.S. Postal Service?

A2: No. A licensee is not required to use the U.S. Postal Service. It may use other delivery services, such as FedEx or UPS.

Q3: What "other means" of advance notification besides the U.S. Postal Service or a commercial delivery service would be acceptable under 10 CFR 37.77(a)(3)?

A3: A licensee may also fax or e-mail notifications. However, if the licensee uses a facsimile or e-mail for this purpose, it should first verify the correct facsimile number or e-mail address with the recipient. The licensee should also confirm receipt of the message with its intended recipient so that both parties will know if the information has gone to an unauthorized third party. Although an NRC licensee may use any of the delivery methods discussed above, the agency recommends e-mail for NRC notification because its Headquarters operations office is staffed 24 hours a day, 7 days a week, and it will be able to confirm receipt of the notification in short order.

§ 37.77, "Advance Notification of Shipment of Category 1 Quantities of Radioactive Material" (continued)

§ 37.77(b), "Information To Be Furnished in Advance Notification of Shipment"

Each advance notification of shipment of category 1 quantities of radioactive material must contain the following information, if available at the time of notification:

§ 37.77(b)(1)
The name, address, and telephone number of the shipper, carrier, and receiver of the category 1 radioactive material;

§ 37.77(b)(2)
The license numbers of the shipper and receiver;

§ 37.77(b)(3)
A description of the radioactive material contained in the shipment, including the radionuclides and quantity;

§ 37.77(b)(4)
The point of origin of the shipment and the estimated time and date that shipment will commence;

§ 37.77(b)(5)
The estimated time and date that the shipment is expected to enter each State along the route;

§ 37.77(b)(6)
The estimated time and date of arrival of the shipment at the destination; and

§ 37.77(b)(7)
A point of contact, with a telephone number, for current shipment information.

Explanation

This provision lists the information that must be provided in an advance notification of shipment.

Q&As

Q1: What information must be included in an advance notification?

A1: Each licensee is expected to make a good-faith effort to provide all the information required by 10 CFR 37.77(b) available when it makes the initial advance notification.

However, the licensee must provide any information that is not available at the time of the initial notification in a revised notification under 10 CFR 37.77(c)(1) as soon as the information becomes available but before the commencement of the shipment. Annex D to this subpart provides an example of an advance notification template that would satisfy the NRC.

Q2: What if I don't have all the shipment information listed in this subsection? Should I make the notification with the information I have, and provide the outstanding information when I'm able to obtain it?

A2: Yes. A shipping licensee may make the initial notification with the information available at that time. The remaining information must be provided in a revision notice as soon as the information becomes available.

Q3: If I don't have all the shipment information listed in this subsection before the scheduled shipment, can I still ship on the date originally specified in my first advance notification, or must I delay until I can provide all the information?

A3: The licensee may ship the radioactive material on the date originally specified in its first advance notification if the only shipment information still missing is the exact quantity of radioactive material that will be shipped. Prior operating experience has shown that, in some cases, the exact quantity of radionuclides that will be shipped is the information that is most likely unknown until shortly before the shipment. However, before commencement of the shipment, the shipping licensee must provide all of the information required by this subsection, including the quantity of material being shipped.

Q4: Must the "point of contact...for current shipment information" under 10 CFR 37.77(b)(7) be someone accompanying the shipment?

A4: No. The point of contact should, in fact, *not* be someone accompanying the shipment because communication to that person could be lost during an accident or security incident. The point of contact should be someone with access to the information on the shipment that is available at the movement control center, as it is defined in 10 CFR 37.5.

§ 37.77, "Advance Notification of Shipment of Category 1 Quantities of Radioactive Material" (continued)

§ 37.77(c), "Revision Notice"

§ 37.77(c)(1)

The licensee shall provide any information not previously available at the time of the initial notification, as soon as the information becomes available but not later than commencement of the shipment, to the governor of the State or the governor's designee and to the NRC's Director of Nuclear Security, Office of Nuclear Security and Incident Response, U.S. Nuclear Regulatory Commission, Washington, DC 20555.

§ 37.77(c)(2)

A licensee shall promptly notify the governor of the State or the governor's designee of any changes to the information provided in accordance with paragraphs (b) and (c)(1) of this section.

The licensee shall also immediately notify the NRC's Director, Division of Security Policy, Office of Nuclear Security and Incident Response, U.S. Nuclear Regulatory Commission, Washington, DC 20555 of any such changes.

§ 37.77(d), "Cancellation Notice"

Each licensee who cancels a shipment for which advance notification has been sent shall send a cancellation notice to the governor of each State or to the governor's designee previously notified and to the NRC's Director, Division of Security Policy, Office of Nuclear Security and Incident Response. The licensee shall send the cancellation notice before the shipment would have commenced or as soon thereafter as possible. The licensee shall state in the notice that it is a cancellation and identify the advance notification that is being cancelled.

Explanation

The shipping licensee must notify each of the previously notified States and the NRC if the schedule for a shipment is revised or cancelled.

Q&As

Q1: What should a licensee do if the shipment of a category 1 quantity is cancelled?

A1: If the shipment is cancelled, the shipping licensee must notify the States to which the initial advance notice had been sent and the regulatory agency (the NRC or Agreement State agency) with jurisdiction in the State of origin of the shipment.

Q2: What should a licensee do if the shipment schedule of a category 1 quantity is revised?

A2: If the shipment schedule is revised, the shipping licensee must notify the State through which the shipment is passing and the States that the shipment is scheduled to enter for the remainder of its itinerary.

Q3: Under 10 CFR 37.77(c)(2), how "promptly" must a category 1 shipping licensee notify the governor of the State or the governor's designee of any changes in the shipment information provided in the first advance notification or any new information not previously available?

A3: The shipping licensee should notify any affected State's governor or designee without delay. More specifically, assuming that there is no change in the names, addresses, or telephone numbers of the shipper, carrier, or receiving licensee or no change in the shipping or receiving licensees' license numbers, the shipping licensee should notify any affected State's governor or designee as soon as it discovers or is advised of the following changes:

- A change is made in the description of the radioactive material that will be shipped, including the radionuclides or quantity.
- A change is made in the shipment's point of origin or estimated time or date of commencement.

- A change is made in the estimated time or date that the shipment is expected to enter each State along the route. (For example, notification of a 6-hour delay in the estimated entry time should be made; however, notification of a 15-minute delay is not necessary.)
- A change is made in the estimated time or date of arrival of the shipment at the destination. (For example, notification of a 6-hour delay should be made; however, a 15-minute delay is not necessary.)
- A change is made in the name or telephone number of the point of contact for current shipment information.

Notification of any changes in shipment information en route should take place as soon as the driver or other authorized member of the transfer crew determines and informs the movement control center of a change in this information. Prompt notification of significant schedule changes is particularly important if a State plans to provide an escort for the shipment.

Q4: Under 10 CFR 37.77(c)(2), the licensee must also notify the NRC "immediately." Is there a meaningful difference between notifying a governor or governor's designee "promptly" and notifying NRC "immediately"?

A4: No. The NRC considers the words to be interchangeable for the purposes of 10 CFR Part 37.

Q5: What mechanism should a licensee use to notify the States and the NRC of revisions or changes in the shipment information?

A5: The rule does not specify a mechanism for notifying the States and the NRC of revisions or changes in the shipment information. Each licensee therefore may use the method that works best for it. However, for cancellations, the licensee must "send a cancellation notice." Correspondences that are e-mailed, faxed, or written are all acceptable methods. The States and the NRC must receive the cancellation notice before commencement of the shipment or as soon as possible thereafter to provide time for a State to cancel any planned escorts. A telephone call sometimes may be necessary to ensure timely receipt of the notice.

Q6: Why do I have to notify the governors or their designees and the NRC under 10 CFR 37.77(c)(2) if information about the shipment changes during the shipment?

Q6: The notification to governors allows States to be aware of shipments in their jurisdictions, to provide escorts if they consider them necessary, and to respond to any incidents that require a State response. The notification to the NRC provides information that allows the NRC and other Federal agencies to respond in the event of an incident.

Q7: Other than the requirement to notify NRC of any revision or changes to the advance notification, are any other notifications required?

A7: No. However, one voluntary good practice would be helpful—that is, to notify the NRC of the shipment's actual arrival or departure times if these differ from the original advance notification. As noted above in Q5 and A5 for 10 CFR 37.77, the NRC shares information about shipments of category 1 quantities of radioactive material with other Federal Government agencies, such as DOT and DHS. Confirmation of actual shipment departure and arrival dates would enable the NRC staff to provide more accurate data for the tracking and current status of these shipments without having to make assumptions or without having to contact the shipper by telephone to get updates. Notification to the NRC within 24 hours of the actual departure and arrival dates should be sufficient for this purpose.

§ 37.77, "Advance Notification of Shipment of Category 1 Quantities of Radioactive Material" (continued)

§ 37.77(e), "Records"

The licensee shall retain a copy of the advance notification and any revision and cancellation notices as a record for 3 years.

Explanation

This definition is self-explanatory

Q&As

Q1: For purposes of this record retention requirement, what constitutes a "revision" of an advance notification? Do I have to retain records of revisions for typos?

A1: The licensee should retain a record of any notification revision that it considered important enough to send to the NRC and the affected State. For example, the licensee should consider an error in one of the digits for a shipment date or time to require a revision of the initial notification. The licensee therefore would retain this type of revision. However, the licensee would not need to issue a revised notification to correct a misspelling of a common word if the misspelling does not have the potential to cause confusion.

Q2: When does the 3-year clock start for the required retention of an advance notification document?

A2: The licensee should start the document retention period on the date that it sent the notification.

Q3: If I have to cancel a planned shipment by phone to assure that I give the NRC or State official notice as soon as possible, how should I document that phone call?

A3: If the licensee has to provide a revision or cancellation notification by telephone to the NRC or State official, it should record the date and time of the call, the name of the caller, the name of the person called, and the purpose of the call (revision or cancellation.) To meet the requirement to "retain a copy" of these revision and cancellation notices, the licensee should keep this documentation of the conversation in electronic or hardcopy form for 3 years after the date of the call. The NRC encourages the licensee to notify the agency by e-mail.

§ 37.77, "Advance Notification of Shipment of Category 1 Quantities of Radioactive Material" (continued)

§ 37.77(f), "Protection of Information"

State officials, State employees, and other individuals, whether or not licensees of the Commission or an Agreement State, who receive schedule information of the kind specified in § 37.77(b) shall protect that information against unauthorized disclosure as specified in § 73.21 of this chapter.

Explanation

Schedule information for shipments of category 1 quantities of radioactive material is considered to be SGI, and anyone receiving the information must protect it in accordance with 10 CFR 73.21.

Q&As

Q1: What do the information protection requirements in 10 CFR 73.21 specify?

A1: The regulations in 10 CFR 73.21 mostly address general performance requirements for the protection of SGI, including SGI with an SGI-M designation or marking. The regulation in 10 CFR 73.21(a)(ii) specifies that the transportation of source, byproduct, or special nuclear material in greater than, or equal to, category 1 quantities of concern must meet the requirements in 10 CFR 73.23.

Q2: Who is required to protect the schedule information?

A2: Any individual who receives the schedule information or any other SGI or SGI-M must protect it in accordance with the requirements in 10 CFR 73.21 through 10 CFR 73.23.

Q3: What do I need to do, if anything, to ensure that those to whom I provide shipment schedule information comply with Part 73 requirements to protect it?

A3: The NRC recognizes that licensees have no control over the use or misuse of this information by persons whom they are required to notify. Accordingly, the NRC or an affected Agreement State agency will not hold a licensee responsible for the compliance of others. However, the licensee must mark any hardcopy document that it provides in accordance with 10 CFR 37.23(d) and must transmit any electronic information in accordance with 10 CFR 37.23(g) to inform the individual who receives the schedule information or any other SGI or SGI-M that he or she must protect it in accordance with the requirements in 10 CFR 73.21 and 10 CFR 73.23. If the licensee uses a delivery service (e.g., FedEx, DHL Express, or UPS) or other commercial carrier (e.g., EnergySolutions, AATCarriers, or Edlow International) to transport its radioactive material, it should make sure that the carrier understands its responsibility to protect shipment-related information.

§ 37.79, "Requirements for Physical Protection of Category 1 and Category 2 Quantities of Radioactive Material during Shipment"

§ 37.79(a), "Shipments by Road"

§ 37.79(a)(1)

Each licensee who transports, or delivers to a carrier for transport, in a single shipment, a category 1 quantity of radioactive material shall:

§ 37.79(a)(1)(i)

Ensure that movement control centers are established that maintain position information from a remote location. These control centers must monitor shipments 24 hours a day, 7 days a week, and have the ability to communicate immediately, in an emergency, with the appropriate law enforcement agencies.

§ 37.79(a)(1)(ii)

Ensure that redundant communications are established that allow the transport to contact the escort vehicle (when used) and movement control center at all times. Redundant communications may not be subject to the same interference factors as the primary communication.

§ 37.79(a)(1)(iii)

Ensure that shipments are continuously and actively monitored by a telemetric position monitoring system or an alternative tracking system reporting to a movement control center. A movement control center must provide positive confirmation of the location, status, and control over the shipment. The movement control center must be prepared to promptly implement preplanned procedures in response to deviations from the authorized route or a notification of actual, attempted, or suspicious activities related to the theft, loss, or diversion of a shipment. These procedures will include, but not be limited to, the identification of, and contact information for, the appropriate LLEA along the shipment route.

§ 37.79(a)(1)(iv)

Provide an individual to accompany the driver for those highway shipments with a driving time period greater than the maximum number of allowable hours of service in a 24-hour duty day as established by the Department of Transportation Federal Motor Carrier Safety Administration. The accompanying individual may be another driver.

Explanation

These provisions establish the security requirements for shipping category 1 quantities of radioactive material by road.

Q&As

Q1: When is the use of a movement control center necessary under 10 CFR 37.79(a)(1)?

A1: To meet the requirements in 10 CFR 37.79(a)(1)(i), any licensee that ships category 1 quantities of radioactive material by road must either establish or use a carrier that has established movement control centers that maintain position information from a location remote from the transport vehicle or trailer. The control centers would be required to monitor shipments on a continuous and active basis 24 hours a day, 7 days a week, and must have the ability to communicate immediately in an emergency with the appropriate law enforcement agencies. The movement control center must provide positive confirmation of the location of the shipment, its status, and the individuals who are in control of the shipment, and it must be prepared to implement preplanned procedures in response to deviations from the authorized route or to a notification of actual or attempted theft or diversion or suspicious activities related to the theft, loss, or diversion of a shipment. These procedures include the identification of, and contact information for, the appropriate LLEA along the shipment route.

Q2: What is meant by "active" monitoring?

A2: A movement control center is monitoring on an active basis whenever it employs a method for tracking a shipment that provides the capability for the control center operators to be immediately aware if a shipment has deviated from the shipping plans. For example, the movement control center must have the capability to be immediately aware if (1) the shipment deviates from the planned route, (2) any unscheduled stops occur, or (3) any scheduled stops take longer than expected.

Q3: Where can the movement control center be located? Does it need to be on the licensee's property?

A3: The movement control center may be located at either the licensee's or a third party's site; however, it must be a stationary facility, not a mobile vehicle. Regardless of location, the movement control center must be able to monitor the category 1 shipment at all times and to communicate with appropriate law enforcement agencies should the need arise.

Q4: Why are redundant communications necessary, and what must these redundant devices or systems be able to do under 10 CFR 37.79(a)(1)(ii)?

A4: Redundant communications must mitigate an interruption caused by either natural events, such as storms, or deliberate actions, such as signal jamming, that may cause communications to be lost on the primary communication device. One or more additional communication devices must be available to operate independently from the primary device, thereby minimizing the possibility that whatever disabled the primary device will also affect the redundant devices. Redundant communications must be in place at all times to allow the transport vehicle to contact the movement control center and nearest LLEA if assistance is needed. An escort vehicle is not required. If the licensee uses an escort vehicle, however, it must also be able to communicate with the transport driver and the movement control center. The redundant communication method must not be subject to the same interference factors as the primary communication method. To meet this requirement, the licensee must ensure that the two systems do not rely on the same hardware or software (e.g., cell tower or proprietary network) to transmit their signal.

Q5: What is a "telemetric position monitoring system" under 10 CFR 37.79(a)(1)(iii)?

A5: A telemetric position monitoring system is a data transfer system that captures, by instrumentation or other measuring devices, information about the location and status of a transport vehicle or package between the departure and destination locations. The gathering of this information permits the remote monitoring and reporting of the location of a transport vehicle or package. Systems that use radiofrequency identification or satellite-based global positioning technologies are examples of telemetric position monitoring systems.

Q6: Is a global positioning system (GPS) required?

A6: No. For category 1 material, continuous and active monitoring of shipments is required without regard to the technology that provides these capabilities. Continuous and active monitoring means that at any time while the shipment is enroute, the licensee must know the shipment's whereabouts. Not specifying a particular technology provides licensees flexibility to design a continuous and active monitoring system that meets their unique circumstances. A GPS would be considered an acceptable method.

Q7: When is a second individual needed for the shipment under § 37.79(a)(1)(iv)?

A7: The licensee must provide an accompanying individual for the shipment when the driving time period is greater than the maximum number of allowable hours of service in a 24-hour duty day, as established by the DOT Federal Motor Carrier Safety Administration. (See 49 CFR Part 395, "Hours of Service of Drivers.") This security measure provides reasonable assurance that the material will be protected from theft or diversion when it is stationary and in emergency situations in which the driver must stop or leave the vehicle. The accompanying individual may be another driver.

Q8: What should the driver and a second accompanying individual do during the shipment of a category 1 quantity?

A8: The driver or the accompanying individual, or both, should periodically call the movement control center to provide a verbal status of the shipment and delivery. Each individual should maintain constant visual surveillance of the surrounding environment during transport. If the driver requires a break and stops the transport vehicle, either the driver or the accompanying individual should maintain constant visual surveillance of the immediate environment of the transport vehicle while it is not in motion. At least one of the individuals should periodically walk around the transport vehicle while it is not in motion to help confirm that there are no apparent safety- or security-related issues associated with the vehicle. In addition, the periodic walkaround should include visual surveillance of the surrounding area to confirm that there is no evidence of tampering with the contents of the vehicle or no unusual or suspicious activity in the immediate vicinity. Note that the accompanying individual should undertake the communication and surveillance measures discussed above if the driver is sleeping during the break.

§ 37.79, "Requirements for Physical Protection of Category 1 and Category 2 Quantities of Radioactive Material during Shipment" (continued)

§ 37.79(a)(1), "Shipments by Road" (continued)

§ 37.79(a)(1)(v)
Develop written normal and contingency procedures to address:

§ 37.79(a)(1)(v)(A)
Notifications to the communication center and law enforcement agencies;

§ 37.79(a)(1)(v)(B)
Communication protocols. Communication protocols must include a strategy for the use of authentication codes and duress codes and provisions for refueling or other stops, detours, and locations where communication is expected to be temporarily lost;

§ 37.79(a)(1)(v)(C)
Loss of communications; and

§ 37.79(a)(1)(v)(D)
Responses to an actual or attempted theft or diversion of a shipment.

§ 37.79(a)(1)(vi)
Each licensee who makes arrangements for the shipment of category 1 quantities of radioactive material shall ensure that drivers, accompanying personnel, and movement control center personnel have access to the normal and contingency procedures.

Explanation

These provisions require the use of written normal and contingency procedures for shipping category 1 quantities of radioactive material by road.

Q&As

Q1: Under 10 CFR 37.79(a)(1)(v), each licensee that makes arrangements for the shipment of category 1 quantities of radioactive material must develop written "normal" and "contingency" procedures. How do I decide what I should address in my "normal" procedures and what should go into my "contingency" procedures?

A1: Normal operating procedures should describe activities conducted to meet regulatory requirements that would be applicable under expected routine operating conditions. Examples of such activities include refueling and comfort stops, meal stops, and a routine check-in.

Contingency procedures should identify issues that could interfere with compliance during the preparation for transport or during the actual transport of the radioactive material. Examples of such interferences include bad weather, suspicious activities, mechanical breakdown, road or bridge closures, detours, accidents, and acute illness. Contingency procedures should address appropriate actions for both anticipated aberrant situations and those that licensees previously encountered during the transport of radioactive material or storage incident to transport.

Q2: 10 CFR 37.79(a)(1)(v)(A) requires my procedures for category 1 shipments to address "notifications to the communication center and law enforcement agencies." What kind of events must I develop notifications for, and what means of communications must I use (i.e., satellite phone, cell phone, e-mail, text messaging, or two-way radio)?

A2: The licensee's procedures must address notifications for all normal and contingency conditions covered by the procedures. The licensee decides on its choice of communications technology. However, the licensee should bear in mind that e-mail and text messaging may not be conducive to communication without delay in the event of an emergency.

Q3: 10 CFR 37.79(a)(1)(v)(B) requires me to develop communication protocols for category 1 shipments that, among other things, include "a strategy for the use of authentication codes and duress codes and provisions for refueling or other stops, detours, and locations where communication is expected to be temporarily lost." What are acceptable methods for authentication and duress codes and acceptable technologies for applying them?

A3: An acceptable protocol may use a number of methods and technologies or combinations of them for duress and authentication codes. The rule does not specify which methods and technologies to use because the optimum strategy may differ from licensee to licensee. However, in choosing a strategy for authentication and duress codes, a licensee should consider the purpose of each. The purpose of an authentication code is to enable a licensee to confirm that the radioactive material remains in the physical possession of an authorized employee of the licensee or carrier or an authorized representative of the NRC, an Agreement State agency, or an LLEA. The purpose of a duress code is to enable a licensee to confirm that the individual at the offsite location who initiated the communication or who is responding to the licensee's or communication center's call is not being forced to provide false information.

For example, to frustrate any attempt by an unauthorized individual to delay an investigation or to call off an ongoing recovery effort prematurely, the licensee should be able to confirm through its preestablished authentication protocol the true identity of the employee, regulatory agency representative, or law enforcement officer reporting from an offsite location. This authentication may be accomplished by using an agreed upon separate radio frequency or alternative communication method, by asking the caller to appear before a video camera on the vehicle to display a photo identification badge, by asking for the correct answer to one or more agreed upon questions, or by using a combination of these methods.

Questions should require specific responses that are either of a personally distinguishing nature (e.g., the name of a first pet) or otherwise not so intuitively obvious that an adversary could infer the correct response. Similarly, the licensee or carrier may use one or several of these techniques in combination in a preestablished protocol or code word or phrase to signal that the driver or accompanying individual is under duress (e.g., at gunpoint or within lethal range of an explosive.) The duress code should permit the driver or accompanying individual to introduce the code on his or her own initiative without prompting, and it may involve seemingly mission-related technical questions and answers, apparently offhand remarks, or some other conversational technique. The purpose of the duress code should be to enable the offsite individual to signal without arousing suspicion that he or she is making a false report under threat by an adversary who is not visible or who is obviously malevolent to licensee or call center personnel.

Q4: Under 10 CFR 37.79(a)(1)(v)(C), my communications protocols for category 1 shipments must address "loss of communications." What constitutes such a "loss of communications"? A loss of a minute's duration? A loss of a half-hour? One dropped call? Two or more dropped calls? Would static be considered "a loss of communication"?

A4: The NRC considers a loss of communication to be any curtailment in the availability of, impairment in the timeliness of, or degradation in the clarity of a message such that the intended recipient of the message is unable to understand it or to receive it when needed. Only a shipping or receiving licensee, a movement control center operator, or a transport crew can determine what constitutes a loss of communications in the context of the importance and urgency of the message for safe and secure delivery of the material. Therefore, the rule leaves this determination to those who are best able to make such a determination under the operating conditions that prevail during the shipment. These conditions necessarily frame any decision on when to switch from the primary to an alternate communications system.

To determine what should be considered a loss of communication, the licensee should identify foreseeable normal and contingency scenarios and should consider their likely urgency in deciding how many times the communications center, driver, or escort should try to restore the primary means of communication before resorting to the secondary means. For example, in an accident or a security emergency, there should be no delay in resorting to the secondary means of communication. The protocol should also allow the communicator to take the severity of static into account and the perceived urgency of a situation in determining the extent to which the degradation of communication clarity justifies switching to the alternate communication technology.

Q5: As a shipper of category 1 quantities under 10 CFR 37.79(a)(1)(vi), I have to “ensure” that drivers, accompanying personnel, and movement control center personnel have access to the normal and contingency procedures. How do I “ensure” that drivers, transport crew members, and movement control center staff have adequate access? If I give them hardcopies of the protocols, am I in violation if one of them loses the hardcopy and forgets to replace it before an inspection?

A5: The NRC understands that limits exist on the extent to which a licensee can “ensure” that drivers, transport crew members, and movement control center staff actually use the licensee’s procedures for normal and contingency conditions. However, a licensee can take certain measures to demonstrate compliance with the requirement to “ensure” that these individuals “have access” to these procedures. Possible measures could include, among others, requiring (in the shipping contract) that the carrier’s driver, the accompanying individual, and the movement control center staff sign a statement before commencement of the shipment that they have in their possession a copy of these procedures and have read them. Another measure could include another clause in the shipping contract with a similar signoff procedure that requires the driver to keep a copy of the procedures in the glovebox or in another readily accessible location inside the vehicle. The licensee could also offer and encourage the training of the carrier’s employees who are responsible for category 1 shipments in the shipping licensee’s procedures. In addition, the shipping licensee could deliver the procedures to the carrier’s movement control center using a commercial delivery service before the first shipment and could retain a receipt for documentation.

§ 37.79, “Requirements for Physical Protection of Category 1 and Category 2 Quantities of Radioactive Material during Shipment” (continued)

§ 37.79(a)(2)

Each licensee that transports category 2 quantities of radioactive material shall maintain constant control and/or surveillance during transit and have the capability for immediate communication to summon appropriate response or assistance.

§ 37.79(a)(3)

Each licensee who delivers to a carrier for transport, in a single shipment, a category 2 quantity of radioactive material shall:

§ 37.79(a)(3)(i)

Use carriers that have established package tracking systems. An established package tracking system is a documented, proven, and reliable system routinely used to transport objects of value. In order for a package tracking system to maintain constant control and/or surveillance, the package tracking system must allow the shipper or transporter to identify when and where the package was last and when it should arrive at the next point of control.

§ 37.79(a)(3)(ii)

Use carriers that maintain constant control and/or surveillance during transit and have the capability for immediate communication to summon appropriate response or assistance; and

§ 37.79(a)(3)(iii)

Use carriers that have established tracking systems that require an authorized signature prior to releasing the package for delivery or return.

Explanation

These provisions establish this part's security requirements for shipping category 2 quantities of radioactive material by road.

Q&As

Q1: I possess a category 2 quantity of radioactive material and transport it without using a commercial carrier service. What are the physical protection requirements for road shipments of these quantities of material?

A1: A licensee that does not use a commercial carrier and that transports by road itself category 2 or greater quantities of material must meet the requirements in 10 CFR 37.79(a)(2). To comply with 10 CFR 37.47(a), the licensee must establish a security zone around the radioactive material and may use the transport vehicle as the security zone boundary. The regulation in 10 CFR 37.47(c) requires the licensee to limit access to the security zone to authorized individuals. In addition, the regulations in 10 CFR 37.49 require the licensee to monitor, detect, assess, and respond to any unauthorized access. The licensee will also need to comply with the preplanning and coordination requirements in 10 CFR 37.75 and with the investigation and reporting requirements in 10 CFR 37.81 if a shipment is lost. For those shipments of category 2 quantities of radioactive material that meet the criteria in 10 CFR 71.97(b) of this chapter, the shipping licensee must also comply with the advance notification provisions of that section.

Q2: What requirements do I have to meet when I deliver a category 2 quantity of radioactive material to a carrier to ship in a single shipment?

A2: The regulation in 10 CFR 37.79(a)(3)(i) requires the licensee to use a carrier that has an established, documented package tracking system that reliably enables it, as the shipping licensee, to know the last location of the package, the time that it was at that location, and the time that it should arrive at the next control point. To meet this requirement, the carrier's tracking system should enable the licensee to see the chain of custody for the package and should show who is accountable for the package at each stage of the trip so that the licensee or the carrier can promptly determine if the shipment is lost or missing. The tracking system must also require an authorized signature before the release of the package for delivery or return. The regulation in 10 CFR 37.79(a)(3)(iii) requires the licensee's carrier to maintain constant control and surveillance during transit and to have the capability to summon an armed LLEA response or other emergency or urgent assistance immediately when necessary.

Q3: If I'm not using an established nationwide delivery system like DHL Express, FedEx, or UPS, what can I do to ensure that my carrier has the equivalent tracking capabilities?

A3: The licensee should ascertain in advance if the carrier maintains a package-tracking system with a reliable on-demand capability to ascertain the last location and current status of the shipment. If the licensee's carrier does not have such a capability, the licensee must either require it as a condition for shipment or find another carrier that meets the requirements in 10 CFR 37.79(a)(2) and 10 CFR 37.79(a)(3) and meets other applicable provisions of this part.

Q4: What must a package control system for a category 2 shipment be able to do to "maintain constant control and/or surveillance"? Would I be in violation if there are any lapses or interruptions?

A4: The regulation in 10 CFR 37.79(a)(3)(i) clarifies the performance requirements for "constant control and/or surveillance" by stating that, for a package tracking system to maintain these capabilities, the system "must allow the shipping licensee or transporter to identify when and where the package was last and when it should arrive at the next point of control." Thus, although the licensee should take all reasonable measures to prevent or minimize tracking system malfunctions, it will not be in violation if the carrier's package control system is not able to pinpoint the location of the shipment moment by moment in real time.

Q5: 10 CFR 37.79(a)(2) requires me to maintain constant control and/or surveillance "during transit." When does "transit" begin and end for the purposes of this requirement?

A5: In general, transit begins when the organization with physical control over the material begins to move it off site. Unless the licensee itself takes the material off site to the carrier for shipment, transit begins when the carrier accepts the consignment of radioactive material for shipment and begins moving the loaded transport vehicle. Transit ends when the receiving licensee accepts the shipment from the carrier and unloads or allows the radioactive material to be unloaded at the agreed upon destination.

Q6: What kind of equipment must I use to "have the capability for immediate communication" under 10 CFR 37.79(a)(2) to summon help? Satellite phones? Cell phones? Laptops for e-mails? Text messaging? Two-way radios? How "immediate" must this communication be?

A6: The rule does not specify any particular communications technology. That choice is left to the licensee's discretion. However, the licensee or carrier must ensure that its chosen technology permits "immediate communication to summon appropriate response or assistance," the same as under 10 CFR 37.79(a)(3)(ii). Cell or satellite phones and two-way radios meet this requirement. The use of e-mail or text messaging does not meet the requirement because the licensee or carrier cannot always be certain that the message has been received and because the message could remain unopened too long to enable an immediate response.

Q7: Do I have to have a backup communications system to maintain the capability for "immediate communication"?

A7: A backup system is not required for transfers of category 2 quantity shipments. However, such a system could provide a greater degree of assurance that the licensee or carrier will be able to maintain its capability for immediate communication in the event of a malfunction or damage to the primary communications system.

Q8: What's the difference between "response" and "assistance"?

A8: The regulations in 10 CFR Part 37 generally use the word "response" to denote a response by law enforcement personnel to an apparent or actual security event or a response by trained emergency services personnel to prevent or mitigate any collateral safety impacts of a security event. The regulations in 10 CFR Part 37 use "assistance" to denote other kinds of emergency or urgent conditions under which the transport crew or vehicle(s) requires outside help, such as towing or roadside repair service, to avoid or minimize an unplanned delay in the shipment. The NRC has used both words in 10 CFR 37.79(a)(2) and 10 CFR 37.79(a)(3) to signify that the licensee or carrier must have the capability for immediate communications for both security- and nonsecurity-related incidents.

Q9: Under 10 CFR 37.79(a)(3)(iii), licensees must use carriers with tracking systems that require an "authorized signature" before the package can be released for delivery or return.

How will I know that an individual who signs the shipping papers for a package is "authorized"?

A9: The individual should be an employee or contractor of the receiving licensee. Because that individual is subject to reassignment and may be unknown to the licensee, the licensee, as good practice, should request that the carrier ask for Government- or licensee-issued photo identification to verify the individual's identity and company-issued documentation to verify that the individual works for the receiving licensee or its contractor.

§ 37.79, "Requirements for Physical Protection of Category 1 and Category 2 Quantities of Radioactive Material during Shipment" (continued)

§ 37.79(b), "Shipments by Rail"

§ 37.79(b)(1)

Each licensee who transports, or delivers to a carrier for transport, in a single shipment, a category 1 quantity of radioactive material shall:

§ 37.79(b)(1)(i)

Ensure that rail shipments are monitored by a telemetric position monitoring system or an alternative tracking system reporting to the licensee, third party, or railroad communications center. The communications center shall provide positive confirmation of the location of the shipment and its status. The communications center shall implement preplanned procedures in response to deviations from the authorized route or to a notification of actual, attempted, or suspicious activities related to the theft or diversion of a shipment. These procedures will include, but not be limited to, the identification of, and contact information for, the appropriate LLEA along the shipment route.

§ 37.79(b)(1)(ii)

Ensure that periodic reports to the communications center are made at preset intervals.

Explanation

These provisions establish the security provisions for shipping category 1 quantities of radioactive material by rail.

Q&As

Q1: What are the physical protection requirements for rail shipments of category 1 quantities of radioactive material?

A1: The regulation in 10 CFR 37.79(b)(1)(i) requires licensees to ensure that rail shipments are monitored either by a telemetric position monitoring system or an alternative tracking system that meets certain criteria for reporting to a licensee, a third party, or a railroad communications center. The communications center must provide positive confirmation of the shipment's status and location. Each center must also be prepared to implement preplanned procedures to respond to deviations from the authorized route, a notification of an actual or attempted theft or diversion of a shipment, or any suspicious activity near the shipment. The procedures must also identify and provide contact information for the appropriate LLEA in each jurisdiction along the shipment route. Rail shipment tracking provides the means for a communications center to immediately report an unusual occurrence that could lead to the theft or diversion of the material. The regulation in 10 CFR 37.79(b)(1)(ii) requires the shipping licensee to ensure that the communications center makes periodic reports at preset intervals.

Q2: What is a telemetric position monitoring system?

A2: A telemetric position monitoring system is a data transfer system that captures, by instrumentation or by other measuring devices, information about the location and status of a transport vehicle or package between the departure and destination locations. The gathering of this information permits the remote monitoring and reporting of the location of a transport vehicle or package. Systems that use radiofrequency identification or satellite-based global positioning technologies are examples of telemetric position monitoring systems.

Q3: What's the difference between a railroad communication center and a movement control center for transport by road?

A3: The two types of centers fulfill equivalent communications functions. The NRC refers to "railroad communication center" because that's the accepted railroad industry term for the railroad-related entity that carries out the functions that correspond to those of a movement control center.

Q4: Also under 10 CFR 37.79(b)(1)(i), the communications center must implement preplanned procedures in response to "deviations" from the authorized route. What constitutes a "deviation"?

A4: A deviation from the authorized route would be any redirection to another rail spur or rail line when the railroad communications center was not aware that the switch was to be made. This could include a relocation of a train within a rail yard if the relocation could adversely affect the ability of the railroad to locate the shipment or to provide required security.

Q5: 10 CFR 37.79(b)(1)(ii) requires licensees to ensure that periodic reports to the communications center are made at preset intervals. What must I do if one of these reports is late? How long should I wait before taking action?

A5: The licensee's shipment contract with the highway carrier or railroad should require its movement control or communications center to notify the licensee immediately if a call from its transport crew is delayed by more than a half of the prearranged reporting interval. The licensee should contact the LLEA immediately if the movement control or communications center cannot establish communications with the transport crew.

§ 37.79, "Requirements for Physical Protection of Category 1 and Category 2 Quantities of Radioactive Material during Shipment" (continued)

§ 37.79(b), "Shipments by Rail" (continued)

§ 37.79(b)(2)

Each licensee who transports, or delivers to a carrier for transport, in a single shipment, a category 2 quantity of radioactive material shall:

§ 37.79(b)(2)(i)

Use carriers that have established package tracking systems. An established package tracking system is a documented, proven, and reliable system routinely used to transport objects of value. In order for a package tracking system to maintain constant control and/or surveillance, the package tracking system must allow the shipper or transporter to identify when and where the package was last and when it should arrive at the next point of control.

§ 37.79(b)(2)(ii)

Use carriers that maintain constant control and/or surveillance during transit and have the capability for immediate communication to summon appropriate response or assistance; and

§ 37.79(b)(2)(iii)

Use carriers that have established tracking systems that require an authorized signature prior to releasing the package for delivery or return.

Explanation

This requirement establishes the security provisions for shipping category 2 quantities of radioactive material by rail.

Q&As

Q1: What are the physical protection requirements for rail shipments of category 2 quantities of radioactive material?

A1: The regulation in 10 CFR 37.79(b)(2)(i) requires a licensee that is shipping category 2 quantities of radioactive material by rail to have, at minimum, the capability to contact the shipping carrier and to determine the approximate location of the shipment. Such licensees must also use a carrier that has a documented, proven, and reliable package tracking system that allows the shipping licensee or carrier to identify when and where the package was when the train last reported and when it should arrive at the next point of control. The regulation in 10 CFR 37.79(b)(2)(ii) requires the carrier to maintain constant control and surveillance during transit and to have the capability for immediate communication to summon an appropriate response or assistance. In addition, the carrier must, under 10 CFR 37.79(b)(2)(iii), require an authorized signature before releasing the package for delivery or return.

Q2: What must a package control system for a category 2 shipment be able to do to "maintain constant control and/or surveillance"? Would I be in violation if there are any lapses or interruptions?

A2: The regulation in 10 CFR 37.79(b)(2)(i) clarifies the performance requirements for "constant control and/or surveillance" by stating that, to maintain these capabilities, the package tracking system must allow the shipping licensee or carrier "to identify when and where the package was last and when it should arrive at the next point of control." Thus, although the shipping licensee should take all reasonable measures to prevent or minimize tracking system malfunctions, the licensee will not be in violation if it is unable to pinpoint the location of the shipment moment by moment in real time.

Q3: Section 37.79(b)(2)(ii) requires me to maintain constant control and/or surveillance "during transit." When does "transit" begin and end for the purposes of this requirement?

A3: Transit begins when the carrier accepts the consignment of radioactive material for shipment electronically or in writing and begins to move the loaded transport vehicle. Transit ends when the receiving licensee accepts the shipment from the carrier electronically or in writing and unloads or allows the radioactive material to be unloaded at the agreed upon destination.

Q4: What kind of equipment must a carrier have to "have the capability for immediate communication" to summon help? Laptops for e-mails? Text messaging? Two-way radios? How "immediate" must this communication be?

A4: The rule does not require any particular communications technology. The carrier has the discretion to choose the type of communication equipment that it will use to summon help under 10 CFR 37.79(b)(2). However, the carrier's chosen technology must permit communication without delay. Cell or satellite phones and two-way radios meet this requirement. The use of e-mail or text messaging does not meet this requirement because the carrier cannot always be certain that the message has been received and because the message could remain unopened too long to enable an immediate response.

Q5: Does my carrier have to have a backup communications system to maintain the capability for "immediate communication"?

A5: A backup system is not required; however, such a system could provide a greater degree of assurance that the carrier will be able to maintain its capability for immediate communication in the event of a malfunction or damage to the primary communications system.

Q6: What's the difference between "response" and "assistance"?

A6: The regulation in 10 CFR Part 37 generally uses the word "response" to denote a response by law enforcement personnel to an apparent or actual security event or a response by trained emergency services personnel to prevent or mitigate any collateral safety impacts of a security event. The regulation in 10 CFR Part 37 uses "assistance" to denote a response to other kinds of emergency or urgent conditions under which the transport crew requires outside help, such as towing or on-location repair service, to avoid or minimize an unplanned delay in the shipment. The NRC has used both words in 10 CFR 37.79(b)(2) to signify that the licensee or carrier must have the capability for immediate communications for both security- and nonsecurity-related incidents.

Q7: Under 10 CFR 37.79(b)(2)(iii), licensees must use carriers with tracking systems that require "an authorized signature" before the package can be released for delivery or return. How will I know that an individual who signs the shipping papers for a package is "authorized"?

A7: The individual should be an employee or contractor of the receiving licensee. Because that individual is subject to reassignment and may be unknown to the licensee, the licensee, as good practice, should request that the carrier or railroad ask for Government- or licensee-issued photo identification to verify the individual's identity and company-issued documentation to verify that the individual works for the receiving licensee or its contractor.

§ 37.79, "Requirements for Physical Protection of Category 1 and Category 2 Quantities of Radioactive Material during Shipment" (continued)

§ 37.79(c), "Investigations"

Each licensee that makes arrangements for the shipment of category 1 quantities of radioactive material shall immediately conduct an investigation upon the discovery that a category 1 shipment is lost or missing.

> Each licensee that makes arrangements for the shipment of category 2 quantities of radioactive material shall immediately conduct an investigation, in coordination with the receiving licensee, of any shipment that has not arrived by the designated no-later-than arrival time.

Explanation

The shipping licensee must investigate immediately if a category 1 quantity shipment is lost or missing or if a category 2 quantity shipment does not arrive by the NLT arrival time.

Q&As

Q1: How should I decide that a shipment of my category 1 quantity of radioactive material is lost or missing?

A1: If the carrier's telemetric position monitoring system or the railroad's communications center cannot tell where a licensee's shipment is within a few minutes after the licensee asks for a status of its shipment, and if the licensee is not confident that the tracking and communications systems are functioning normally, it should consider the shipment lost or missing. Additionally, if the licensee cannot reach the transport crew after a missed check-in, it should assume that the shipment is missing.

Q2: What should an investigation of a lost or missing category 1 quantity shipment include?

A2: The scope of an investigation will depend on the circumstances of the lost or unaccounted for material, the shipment's transportation mode, and the contingency procedures. The licensee should coordinate its investigation with the carrier. The investigation should include, but should not be limited to, the following actions:

- Determine the time and location of the last transport crew check-in.
- Determine where communication was lost.
- Determine where tracking was lost.
- Confirm that the equipment is working properly.
- Contact the escort if one was being used.

Q3: What should I do if my category 2 shipment does not arrive by the NLT arrival time?

A3: If the licensee is a receiving licensee, it must notify the shipping licensee in accordance with 10 CFR 37.75(c). In addition, the receiving licensee should cooperate in the investigation that the shipping licensee is required to conduct. If the licensee is the shipping licensee, it must immediately begin an investigation to determine what happened to the shipment.

Q4: What should an investigation of a lost or missing category 2 quantity shipment include?

A4: The shipping licensee should contact the carrier to determine the shipment's last known location. The shipping licensee should work with the carrier to determine the shipment's current location. If the carrier cannot determine the location of the shipment, the shipping licensee should contact the NRC Operations Center to inform it that a category 2 shipment is lost or missing. The shipper should continue to work with the carrier to locate the shipment; however, if the shipment is still missing after 24 hours of the initial notification, the licensee again must contact the NRC Operations Center.

§ 37.81, "Reporting of Events"

§ 37.81(a)

The shipping licensee shall notify the appropriate LLEA and the NRC Operations Center ((301) 816-5100) within 1 hour of its determination that a shipment of category 1 quantities of radioactive material is lost or missing. The appropriate LLEA would be the law enforcement agency in the area of the shipment's last confirmed location. During the investigation required by § 37.79(c), the shipping licensee will provide agreed upon updates to the NRC Operations Center on the status of the investigation.

Explanation

The shipping licensee must notify the LLEA and the NRC Operations Center within 1 hour after determining that a shipment of category 1 quantities of radioactive material is lost or missing.

Q&As

Q1: When must a licensee make notification that a category 1 shipment is lost or missing?

A1: When a licensee determines that a shipment of a category 1 quantity of radioactive material is lost or missing, it must notify the LLEA in the area of the shipment's last confirmed location within 1 hour and must then notify the NRC Operations Center. Notification to the NRC should be as prompt as possible, but not at the expense of causing a delay or interference with the LLEA response to the event. Although 10 CFR 37.81(a) sets 1 hour as the maximum time for a "lost or missing" notification of a category 1 quantity shipment, the licensee should notify the LLEA as soon as the carrier has completed its first unsuccessful attempt to locate the material and has confirmed that its inability to trace it was not a result of human error or a malfunction of the system for monitoring the position of the shipment.

Q2: How can I know when a category 1 shipment is lost or missing?

A2: The regulations in 10 CFR 37.79(a) and 10 CFR 37.79(b) require the licensee to ensure that shipments are continuously and actively monitored by a telemetric position monitoring system or an alternative tracking system that reports to a movement control center for road shipments or a railroad communication center. In addition, the regulations in 10 CFR 37.79(a) and 10 CFR 37.79(b) require the licensee to ensure that both types of centers provide positive confirmation of the location, status, and control over the shipment.

Thus, if the licensee complies with these requirements, it should have no trouble in readily ascertaining if a category 1 shipment is lost or missing. If the transport crew has not checked in at the required frequency or if the tracking system indicates that the shipment is not on the approved route and the licensee has not heard from the transport crew, the licensee should begin an immediate investigation to determine what is happening. If the licensee's attempts to contact the transport crew are unsuccessful (using both the primary and backup communication systems), the licensee should assume that something is wrong and should report to the LLEA and the NRC. In addition, if the licensee is shipping a category 1 quantity of material by road and if it wants to enhance the speed and authenticity of its carrier's confirmation of a lost or missing shipment, the licensee could request that its carrier adopt one or more of TSA's voluntary security action items for highway security-sensitive materials (HSSM). Appendix B on TSA's Web page lists category 1 and category 2 quantities of radioactive materials from IAEA's Code of Conduct list as Tier I HSSM, which TSA defines as commercial motor vehicle-transported materials "whose potential consequences from an act of terrorism include a highly significant level of adverse effects on human life, environmental damage, transportation system disruption, or economic disruption." The following two voluntary practices for Tier I HSSMs, among others in TSA's Appendix A list of security action items, could facilitate a timely determination that licensed material is lost or missing:

1. Panic Button Capability. Employers should implement means for a driver to transmit an emergency alert notification to dispatch. "Panic button" technology enables a driver to remotely send an emergency alert notification message either through the use of satellite or terrestrial communications or the remote panic button to disable the vehicle.
2. Tractor and Trailer Tracking Systems. Employers should have the ability to implement methods for tracking the tractor and trailer throughout the intended route with satellite- or land-based wireless GPS communications systems or both. Tracking methods for the tractor and trailer should provide current position by latitude and longitude. Geofencing and route-monitoring capabilities allow authorized users to define and monitor routes and risk areas. If the tractor or trailer deviates from a specified route or if it enters a risk area, an alert notification should be sent to the dispatch center. An employer or an authorized representative should have the ability to remotely monitor trailer "connect" and "disconnect" events. Employers or an authorized representative should have the ability to poll the tractor and trailer tracking units to request a current location and status report. The tractor position reporting frequency should be configured at not more than 15-minute intervals.

These voluntary practices would also facilitate timely notifications under this section.

Q3: How frequently should the licensee provide updates to the NRC Operations Center?

A3: The licensee should provide updates when it receives new information. In addition, the licensee should discuss with the NRC, on a case-by-case basis, what the frequency of updates should be because the frequency will depend on the circumstances of the lost or missing shipment.

§ 37.81, "Reporting of Events" (continued)

§ 37.81(b)

The shipping licensee shall notify the NRC Operations Center ((301) 816-5100) within 4 hours of its determination that a shipment of category 2 quantities of radioactive material is lost or missing. If, after 24 hours of its determination that the shipment is lost or missing, the radioactive material has not been located and secured, the licensee shall immediately notify the NRC Operations Center.

Explanation

The shipping licensee must notify the NRC Operations Center within 4 hours of the determination that a shipment of category 2 quantities of radioactive material is lost or missing and must call the center after 24 hours if the shipment has not been located.

Q&As

Q1: When must a licensee make notification that a category 2 shipment is lost or missing?

A1: The concept of lost or missing licensed material is neither new nor unique to 10 CFR Part 37. In summary, if the licensee does not know where the material is and cannot readily locate it, it should consider the material lost or missing. The licensee is allowed reasonable time to attempt to trace a shipment. A licensee is not required to make the notification when a shipment has not arrived by the NLT arrival time. The shipping licensee is required to begin its investigation at that point in time. If the investigation does not discover the location of the material, the licensee must notify the NRC. A licensee can take a reasonable amount of time to conduct the investigation; however, the investigation should not last several days. The licensee needs to use some judgment in making the determination that the investigation has taken long enough and that the NRC needs to be notified. Once the licensee determines that the shipment is lost or missing, it must notify the NRC Operations Center within 4 hours of such a determination. In addition, the licensee must immediately notify the NRC Operations Center if, after 24 hours from its determination that the shipment was lost or missing, it still cannot determine the location of the material.

§ 37.81, "Reporting of Events" (continued)

§ 37.81(c)

The shipping licensee shall notify the designated LLEA along the shipment route as soon as possible upon discovery of any actual or attempted theft or diversion of a shipment or suspicious activities related to the theft or diversion of a shipment of a category 1 quantity of radioactive material. As soon as possible after notifying the LLEA, the licensee shall notify the NRC Operations Center ((301) 816-5100) upon discovery of any actual or attempted theft or diversion of a shipment or any suspicious activity related to the shipment of category 1 radioactive material.

Explanation

The shipping licensee must notify the designated LLEA along the shipment route upon discovery of any actual or attempted theft or diversion of a shipment or upon discovery of suspicious activities related to a shipment of category 1 quantities of radioactive material. After notifying the LLEA, the licensee must notify the NRC Operations Center.

Q&As

Q1: This subsection requires me to notify an LLEA along the route as soon as possible "upon discovery" of any actual or attempted theft or diversion of a shipment or suspicious activities. Does this mean I must report whatever a member of the transport crew "discovers" as soon as he or she reports it without first trying to confirm whether it is what the individual thinks it is?

A1: No. However, the licensee should immediately assess if the reported event or activity— or the signal that it received from an alarm or monitoring device—requires LLEA assistance without delay. For example, an explosion on or around the transport or escort vehicle, gunshots nearby in an urban or suburban area or other setting where hunting is prohibited, or the image of an armed individual on the licensee's or the carrier's monitoring screen would indicate the need for an immediate LLEA response and should be reported immediately.

Q2: What would a licensee be required to do if there is an attempt to steal or divert a category 1 shipment?

A2: For shipments of category 1 quantities of radioactive material, a licensee that discovers, under 10 CFR 37.81(c), an actual or attempted theft or diversion of a shipment or any suspicious activity related to a shipment must notify the designated LLEA along the shipment route as soon as possible. After notifying the LLEA, the licensee must notify the NRC Operations Center at (301) 816-5100. The NRC Operations Center will notify other affected States and the agency's Federal partners, as appropriate.

Q3: What types of activities should be considered "suspicious"?

A3: TSA guidance for private or contract carrier employees lists a number of activities that may be considered suspicious. The licensee could require its carrier to make its employees aware of the following suspicious activities:

- discovery of any unknown articles found in, on, or around company vehicles
- unauthorized persons photographing company vehicles or locations or both
- carriers being followed by other vehicles for extended periods of time
- carriers being asked inappropriate questions about the licensee's business related to security, products, delivery schedules, or any other information that seems suspicious
- reports or observations of commercial vehicles parked in unusual locations, such as fields, vacant warehouses, or other secluded areas
- unauthorized persons loitering in areas in which company vehicles are loading/unloading, parked, or serviced

TSA guidance for independent or owner-operated truck drivers offers the following related additional examples of suspicious activities:

- anyone who inquires too much about what the carrier is hauling, where it is going, or what is the intended route
- vehicles following the carrier for extended periods of time
- persons observing parking areas for trucks and automobiles repeatedly cruising through those areas
- requests to stop for alleged accidents or breakdowns that cannot be seen
- anyone approaching the carrier seeking a ride for himself or herself or for another person
- an unattended vehicle parked in or on critical infrastructure (e.g., a tunnel or bridge) or in a location that seems out of place
- anyone avoiding a security checkpoint
- obviously altered signage on a vehicle
- missing license plate(s) or other unique markings on a vehicle

For example, if the truck driver notices suspicious activity, he or she should proceed in accordance with the employer's established security policies and procedures. If the driver has any doubt in regard to suspicious behavior, he or she should report to his or her supervisor.

The lists above are not exhaustive; however, they provide examples of actions that could be considered suspicious. The licensee is responsible for evaluating activity to determine if it is suspicious or not.

§ 37.81, "Reporting of Events" (continued)

§ 37.81(d)

The shipping licensee shall notify the NRC Operations Center ((301)-816-5100) as soon as possible upon discovery of any actual or attempted theft or diversion of a shipment, or any suspicious activity related to the shipment, of a category 2 quantity of radioactive material.

Explanation

The shipping licensee must notify the NRC upon discovery of any actual or attempted theft or diversion of a shipment or suspicious activities related to a shipment of a category 2 quantity of radioactive material.

Q&As

Q1: This subsection requires me to notify the NRC as soon as possible "upon discovery" of any actual or attempted theft or diversion, or suspicious activities related to a shipment of a category 2 quantity. Does this mean I must report whatever my carrier has "discovered" as soon as the carrier reports it without first trying to confirm whether it is what the carrier thinks it is?

A1: No. The regulation in 10 CFR 37.81(b) requires the licensee to notify the NRC Operations Center ((301) 816-5100) within 4 hours of its determination that a shipment of category 2 quantities of radioactive material is lost or missing. This timeframe should allow the licensee to gather sufficient information to decide if there actually have been suspicious activities or an actual or attempted theft or diversion.

Q2: What would a licensee be required to do if there is an attempt to steal or divert a category 2 shipment?

A2: The regulation in 10 CFR 37.81(d) requires a licensee that discovers an actual or attempted theft or diversion of a category 2 quantity shipment or any suspicious activity related to a shipment to notify the NRC Operations Center at (301) 816-5100 as soon as possible.

Q3: What types of activities should be considered "suspicious"?

A3: See Q3 and A3 for 10 CFR 37.81(c).

§ 37.81, "Reporting of Events" (continued)

§ 37.81(e)

The shipping licensee shall notify the NRC Operations Center ((301) 816-5100) and the LLEA as soon as possible upon recovery of any lost or missing category 1 quantities of radioactive material.

Explanation

The shipping licensee must notify the NRC and the LLEA upon recovery of any lost or missing category 1 quantities of radioactive material.

Q&As

Q1: What constitutes "recovery" for purposes of this notification?

A1: The licensee can consider lost or missing licensed material recovered only when it is again in the physical possession of, or in a location otherwise under the control of, an authorized employee of the carrier, the shipping or recipient licensee, an authorized State or Federal agency, or an LLEA that is able to prevent or deter unauthorized access to the material.

Q2: How can I know that the individual reporting the recovery of a lost or missing category 1 shipment is the person he or she says she is and not an adversary trying to trick me into calling off my search?

A2: To frustrate any attempt by an unauthorized individual to call off an ongoing recovery effort prematurely, the licensee should try to confirm the true identity of the employee, regulatory agency representative, or law enforcement officer who is reporting from an offsite location. If the vehicle has an onboard video camera, the licensee may accomplish this confirmation by asking the caller to appear before the camera and display a photo identification badge. If the vehicle does not have an onboard video camera, a licensee could confirm the identity of an employee by using its preestablished authentication code, by asking the employee to use an agreed upon separate radiofrequency or alternative communication method, by asking for the correct answer to one or more agreed upon identifying questions, or by using a combination of these. Questions should require specific responses that are either of a personally distinguishing nature (e.g., the name of a first pet) or otherwise not so intuitively obvious that an adversary could infer the correct response.

To confirm the true identity of a regulatory agency representative or law enforcement officer who claims to have recovered the lost or missing material in cases in which the vehicle does not have an onboard video camera to verify this claim, the licensee could ask the individual to describe the location and to wait with the material until additional law enforcement assistance arrives. If the individual says that his or her current location is unsafe, the licensee may ask for the address or distinguishing landmarks of the alternate location and tell the individual to wait for assistance at that location while it directs the nearest operating LLEA to the new location. If the individual claims to be a law enforcement officer, the licensee could ask the responding LLEA to verify the individual's employment or to refer it to a number at which it can obtain such verification. If the individual claims to be a regulatory agency representative, the licensee could ask for the agency's toll free number to verify his or her employment. The individual's responses to the licensee's requests for more information may not provide the positive identification that it needs; however, incorrect or misleading information may signify that the individual is not the official he or she claims to be. If the licensee suspects a false identity, it should try to keep the individual on the phone and silently alert a colleague if possible (e.g., with a handwritten note) to contact the responding LLEA out of the caller's earshot. This might enable the LLEA to work with the phone service provider to identify more accurately the caller's current location and direction of travel.

The licensee or carrier should also use its preestablished duress code to confirm that an offsite interlocutor is not being forced (e.g., at gunpoint or within lethal range of an explosive) to submit a false report. The code should permit the offsite individual to introduce it on his or her own initiative without prompting and may involve seemingly mission-related technical questions and answers, apparently offhand remarks, or some other conversational technique to enable the offsite individual to signal without arousing suspicion that he or she is making a false report to licensee or call center personnel under threat by an adversary who is not visible or who is obviously malevolent. See also Q3 and A3 for 10 CFR 37.79(a)(1)(v)(B).

§ 37.81, "Reporting of Events" (continued)

§ 37.81(f)

The shipping licensee shall notify the NRC Operations Center ((301) 816-5100) as soon as possible upon recovery of any lost or missing category 2 quantities of radioactive material.

Explanation

The shipping licensee must notify the NRC upon recovery of any lost or missing category 2 quantities of radioactive material.

Q&As

Q1: Should licensees make notification that a lost or missing category 2 shipment has been found?

A1: Yes. The licensee must notify the NRC Operations Center when a lost or missing shipment of category 2 quantities of radioactive material has been located. An Agreement State licensee notifies the Agreement State. This notification would be considered an update of the initial notification.

Q2: Can I notify the NRC Operations Center by any other means than telephoning?

A2: Yes. The licensee can fax a notification to the NRC Operations Center at 301-816-5151.

§ 37.81, "Reporting of Events" (continued)

§ 37.81(g)

The initial telephonic notification required by paragraphs (a) through (d) must be followed within a period of 30 days by a written report submitted to the NRC by an appropriate method listed in § 37.7. A written report is not required for notifications on suspicious activities required by paragraphs (c) and (d) of this section. In addition, the licensee shall provide one copy of the written report addressed to the Director, Division of Security Policy, Office of Nuclear Security and Incident Response. The report must set forth the following information:

§ 37.81(g)(1)

A description of the licensed material involved, including kind, quantity, and chemical and physical form;

§ 37.81(g)(2)

A description of the circumstances under which the loss or theft occurred;

§ 37.81(g)(3)

A statement of disposition, or probable disposition, of the licensed material involved;

§ 37.81(g)(4)

Actions that have been taken, or will be taken to recover the material; and

§ 37.81(g)(5)

Procedures or measures that have been, or will be, adopted to ensure against a recurrence of the loss or theft of licensed material.

Explanation

The licensee must submit a written report to the NRC within 30 days of an initial report of lost or missing material or attempted or actual theft or diversion of a shipment of category 1 or category 2 quantities of radioactive material.

Q&As

Q1: What level of detail should I provide in my 30-day written report on an event?

A1: The appropriate level of detail would depend on a number of considerations, including the nature and severity of the security incident, whether it was a first occurrence or a reoccurrence, and whether it has a significant potential to recur, the adequacy of the licensee's monitoring efforts, the timeliness of the initial detection, the accuracy of the assessment, and the timeliness and potential effectiveness of the measures implemented in response to the incident. In addition to an identification of corrective actions, the report should describe the data and analyses that support the licensee's identification of the root and significant contributing cause(s) of the incident and should explain how these data and analyses support the licensee's selection of corrective actions identified. If the incident was a recurrence of a similar incident, the report should briefly describe past corrective actions and the findings of past assessments and should identify likely reasons that the past corrective actions did not prevent or mitigate a recurrence of the condition. The report should then explain the basis for the licensee's determination that the proposed new or revised corrective actions or changes in the licensee's implementation of these actions are likely to reduce the probability of a recurrence of the condition or to mitigate its effects.

§ 37.81, "Reporting of Events" (continued)

§ 37.81(h)

Subsequent to filing the written report, the licensee shall also report any additional substantive information on the loss or theft within 30 days after the licensee learns of such information.

Explanation

After filing the 30-day report, the licensee must report any additional substantive information on the loss or theft within 30 days after it learns of such information.

Q&As

Q1: What would be considered to be "additional substantive information"?

A1: Examples of additional substantive information would include information not previously reported to the NRC that has been gathered since the previous report was made. If the licensee could not complete its investigation within the 30-day limit for the initial report, the subsequent report would need to provide the final findings.

Q2: Would I have to report as "additional substantive information" something I "learned of" through TV, radio, or print news? Through Wikipedia or some other Internet source? From an LLEA representative?

A2: No. The licensee would not necessarily need to report as substantive information something that it learned through television, radio, print news, or Internet sources. The fact that information is available does not mean that the information is correct or that it would be of interest to the NRC. The information provided by an LLEA representative might be considered substantive, depending on the content. The licensee should use judgment in deciding if the information obtained from any source would be of interest to the NRC.

ANNEX D. TEMPLATE FOR ADVANCE NOTIFICATIONS TO THE NRC OF SHIPMENTS OF CATEGORY 1 QUANTITIES OF RADIOACTIVE MATERIAL UNDER 10 CFR 37.77(B)

Notification Date: ______________________________

Shipping Identification (unique identifier): ______________________________

Notification Revision Number: ______________________________

Shipper (include name,* address,* and name of the point of contact and telephone number):*

Shipper's License Number* (and import or export license, when applicable): ________

Type of Shipment (domestic, import, export, or transshipment): ______________

Point of Origin* (include address and name of the point of contact and telephone number):

Recipient (consignee) (include name,* address,* name of the point of contact, and telephone number):

Recipient's License Number:* ______________________

Radioactive Isotope:* ______________________

Estimated Activity* (preferably in terabecquerel): ______________________

Description of Shipment (e.g., physical form and number of sources or flasks or containers):*

End Use of Shipment:

Carrier(s) (include name,* address,* and name of the point of contact, and telephone number*):

Planned Date* and Time* of Shipment Departure and Confirmation of Shipment Actual Departure Date* and Time*: ______________________

Planned Date* and Time* of Shipment Arrival and Confirmation of Shipment Actual Arrival Date* and Time*: ______________________

Routing Information:

Mode(s) of Transportation: ______________________

Estimated Time and Date That the Shipment Is Expected To Enter Each State along the Route*:

For Import/Exports Include Point of U.S. Entry or Departure:* ______________

Point of Contact, with a Telephone Number for Current Shipment Information*:

* This symbol indicates that the information is required by 10 CFR 37.77(b). The other items are voluntary but are requested by the U.S. Nuclear Regulatory Commission (NRC) to enable it to provide more accurate and detailed information for any emergency response.

Note that the preferred notification method to the NRC is through the Operations Center by e-mail or facsimile at least 4 days before the scheduled shipping date. The notification to the NRC may be made by e-mail to RAMQC SHIPMENTS@nrc.gov or by fax to (301) 816-5151. Shippers should coordinate with States as indicated in 10 CFR 37.77(a). The NRC requests that the shipper also notify the agency with a confirmation of the departure date within 1 day after the actual shipment departure and with a confirmation of the arrival date within 1 day after the actual arrival.

Under 10 CFR 37.77(c)(1), any change in the shipment information supplied above must be provided to the States and the NRC as soon as the information becomes available but before the commencement of the shipment. Under 10 CFR 37.77(c)(2), notification of changes to the shipment schedule while the shipment is in transit must be made immediately to the NRC Operations Center and the appropriate States.

Subpart E—Reserved

SUBPART F—RECORDS

§ 37.101, "Form of Records"

§ 37.103, "Record Retention"

§ 37.101, "Form of Records"

§ 37.101

Each record required by this part must be legible throughout the retention period specified by each Commission regulation. The record may be the original or a reproduced copy or a microform, provided that the copy or microform is authenticated by authorized personnel and that the microform is capable of producing a clear copy throughout the required retention period. The record may also be stored in electronic media with the capability for producing legible, accurate, and complete records during the required retention period. Records, such as letters, drawings, and specifications, must include all pertinent information, such as stamps, initials, and signatures. The licensee shall maintain adequate safeguards against tampering with, and loss of, records.

Explanation

Licensees are required to keep records that are legible for the length of the retention period.

Q&As

These provisions are not unique to 10 CFR Part 37; therefore, no questions and answers are provided.

§ 37.103, "Record Retention"

§ 37.103

Licensees shall maintain the records that are required by the regulations in this part for the period specified by the appropriate regulation. If a retention period is not otherwise specified, these records must be retained until the Commission terminates the facility's license. All records related to this part may be destroyed upon Commission termination of the facility license.

Explanation

Licensees are required to maintain records required by the regulations until the Commission terminates the license, unless the regulations specify a retention period.

Q&As

Q1: If 10 CFR Part 37 does not specify how long to keep a record, is the licensee required to keep it? If so, for how long?

A1: If the regulations in 10 CFR Part 37 do not specify the length of record retention for a particular record, the licensee must retain the record under this section until the Commission terminates the facility license.

Subpart G—Enforcement

§ 37.105, "Inspections"

§ 37.107, "Violations"

§ 37.109, "Criminal Penalties"

§ 37.105, "Inspections"

§ 37.105(a)

Each licensee shall afford to the Commission at all reasonable times opportunity to inspect category 1 or category 2 quantities of radioactive material and the premises and facilities wherein the nuclear material is used, produced, or stored.

§ 37.105(b)

Each licensee shall make available to the Commission for inspection, upon reasonable notice, records kept by the licensee pertaining to its receipt, possession, use, acquisition, import, export, or transfer of category 1 or category 2 quantities of radioactive material.

Explanation

The NRC has the right to inspect the licensee's facilities, including all licensee-maintained records that pertain to its receipt, possession, use, acquisition, import, export, or transfer of radioactive material.

Q&As

These provisions are not unique to 10 CFR Part 37; therefore, no questions and answers are provided.

§ 37.107, "Violations"

§ 37.107(a)

The Commission may obtain an injunction or other court order to prevent a violation of the provisions of—

§ 37.107(a)(1)

The AEA, as amended;

§ 37.107(a)(2)

Title II of the Energy Reorganization Act of 1974, as amended; or

§ 37.107(a)(3)

A regulation or order issued pursuant to those Acts.

Explanation

The NRC may obtain a court order to stop someone from violating the regulations.

Q&As

None. The NRC Enforcement Manual is the implementing guidance for the agency's enforcement program. For additional information on the manual and for links to related documents, click on http://www.nrc.gov/aboutnrc/regulatory/enforcement/ guidance.html #manual. The NRC Enforcement Manual provides guidance consistent with the NRC

Enforcement Policy, which establishes the general principles that govern the enforcement program. For additional information on the NRC Enforcement Policy and for links to related documents, click on http://www.nrc.gov/about-nrc/regulatory/ enforcement/enforce-pol.html.

§ 37.107, "Violations" (continued)

§ 37.107(b)
The Commission may obtain a court order for the payment of a civil penalty imposed under Section 234 of the AEA:

§ 37.107(b)(1)
For violations of—

§ 37.107(b)(1)(i)
Sections 53, 57, 62, 63, 81, 82, 101, 103, 104, 107, or 109 of the Atomic Energy Act of 1954, as amended:

§ 37.107(b)(1)(ii)
Section 206 of the Energy Reorganization Act;

§ 37.107(b)(1)(iii)
Any rule, regulation, or order issued pursuant to the sections specified in paragraph (b)(1)(i) of this section;

§ 37.107(b)(1)(iv)
Any term, condition, or limitation of any license issued under the sections specified in paragraph (b)(1)(i) of this section.

§ 37.107(b)(2)
For any violation for which a license may be revoked under Section 186 of the AEA, as amended.

Explanation

The NRC may obtain a court order to force someone to pay a civil penalty for violating the regulations.

Q&As

None. The NRC Enforcement Manual is the implementing guidance for the agency's enforcement program. For additional information on the manual and for links to related documents, click on http://www.nrc.gov/about-nrc/regulatory/enforcement/ guidance.html# manual. The NRC Enforcement Manual provides guidance consistent with the NRC

Enforcement Policy, which establishes the general principles that govern the enforcement program. For additional information on the NRC Enforcement Policy and for links to related documents, click on http://www.nrc.gov/about-nrc/regulatory/ enforcement/enforce-pol.html.

§ 37.109, "Criminal Penalties"

§ 37.109(a)

Section 223 of the AEA, as amended, provides for criminal sanctions for willful violation of, attempted violation of, or conspiracy to violate any regulation issued under Sections 161b, 161i, or 161o of the Act. For purposes of Section 223, all the regulations in Part 37 are issued under one or more of Sections 161b, 161i, or 161o, except for the sections listed in paragraph (b) of this section.

§ 37.109(b)

The regulations in Part 37 that are not issued under Sections 161b, 161i, or 161o for the purposes of Section 223 are as follows: §§ 37.1, 37.3, 37.5, 37.7, 37.9, 37.11, 37.13, 37.107, and 37.109.

Explanation

Anyone who willfully violates, attempts to violate, or conspires to violate the regulations can be criminally prosecuted.

Q&As

None. The NRC Enforcement Manual is the implementing guidance for the agency's enforcement program. For additional information on the manual and for links to related documents, click on http://www.nrc.gov/about-nrc/regulatory/enforcement/ guidance.html# manual. The NRC Enforcement Manual provides guidance consistent with the NRC Enforcement Policy, which establishes the general principles that govern the enforcement program. For additional information on the NRC Enforcement Policy and for links to related documents, click on http://www.nrc.gov/about-nrc/regulatory/ enforcement/enforce-pol.html.

Appendix A. Category 1 and Category 2 Radioactive Materials

Radioactive Material Thresholds

Table A-1 provides the thresholds for Category 1 and Category 2 quantities of radioactive material.

Table A-1. Category 1 and Category 2 Radioactive Material Thresholds*

RADIOACTIVE MATERIAL	CATEGORY 1 (TBq)**	CATEGORY 1 (Ci)	CATEGORY 2 (TBq)	CATEGORY 2 (Ci)
Americium-241	60	1,620	0.6	16.2
Americium-241/Beryllium	60	1,620	0.6	16.2
Californium-252	20	540	0.2	5.40
Cobalt-60	30	810	0.3	8.10
Curium-244	50	1,350	0.5	13.5
Cesium-137	100	2,700	1	27.0
Gadolinium-153	1,000	27,000	10	270
Iridium-192	80	2,160	0.8	21.6
Plutonium-238	60	1,620	0.6	16.2
Plutonium-239/Beryllium	60	1,620	0.6	16.2
Promethium-147	40,000	1,080,000	400	10,800
Radium-226	40	1,080	0.4	10.8
Selenium-75	200	5,400	2	54.0
Srontium-90	1,000	27,000	10	270
Thulium-170	20,000	540,000	200	5,400
Ytterbium-169	300	8,100	3	81.0

*See the discussion below under "Calculations concerning Multiple Sources or Multiple Radionuclides."

**The terabecquerel values are the regulatory standard. The curie values specified are obtained by converting from the terabecquerel values. The curie values are provided for practical usefulness only.

Calculations concerning Multiple Sources or Multiple Radionuclides

The licensee must use the "sum-of-fractions" methodology to evaluate combinations of multiple sources or multiple radionuclides when determining if a location meets or exceeds the threshold and is thus subject to the requirements in Title 10 of the *Code of Federal Regulations* (10 CFR) Part 37, "Physical Protection of Category 1 and Category 2 Quantities of Radioactive Material."

I. If multiple sources of the same radionuclide or multiple radionuclides are aggregated at a location, the licensee must determine the sum of the ratios of the total activity of each of the radionuclides to verify whether the activity at the location is less than the category 1 or category 2 thresholds in Table A-1, as appropriate. If the calculated sum of the ratios, using the equation below, is greater than or equal to 1.0, the applicable requirements of this part apply.

II. First, determine the total activity for each radionuclide from Table A-1. This determination is done by adding the activity of each individual source, the material in any device, and any loose or bulk material that contains the radionuclide. Then, use the equation below to calculate the sum of the ratios by inserting the total activity of the applicable radionuclides from Table A-1 in the numerator of the equation and the corresponding threshold activity from Table A-1 in the denominator of the equation. The licensee must perform the calculations in metric values (i.e., terabecquerel); the numerator and denominator values must be in the same units.

$$\sum_{1}^{n}\left[\frac{R_1}{AR_1}+\frac{R_2}{AR_2}+\frac{R_n}{AR_n}\right]\geq 1.0\,,$$

where:

R1 = total activity for radionuclide 1

R2 = total activity for radionuclide 2 RN = total activity for radionuclide n

AR1 = activity threshold for radionuclide 1

AR2 = activity threshold for radionuclide 2 ARN = activity threshold for radionuclide n

Explanation

Appendix A establishes the thresholds for category 1 and category 2 radioactive materials.

Q&As

See the Q&As on the definition of category 1 and category 2 radioactive materials (10 CFR 37.5).

INDEX

#

A

B

C

D

E

F

J

K

L

M

N

O

P

Q

R

S

T

U

V

W

Y